■ SECOND EDIT

STRETCHING ANATOMY

Arnold G. Nelson

Jouko Kokkonen

Human Kinetics

Library of Congress Cataloging-in-Publication Data

Nelson, Arnold G., 1953-
  Stretching anatomy / Arnold G. Nelson, Jouko Kokkonen. -- Second edition.
     pages cm
1. Muscles--Anatomy. 2. Stretch (Physiology) I. Kokkonen, Jouko. II. Title.
QM151.N45 2014
611'.73--dc23
                          2013013541
ISBN: 978-1-4504-3815-5 (print)

Acquisitions Editor: Tom Heine; **Developmental Editor:** Cynthia McEntire; **Assistant Editor:** Elizabeth Evans; **Copyeditor:** Patricia MacDonald; **Graphic Designer:** Fred Starbird; **Graphic Artist:** Julie L. Denzer; **Cover Designer:** Keith Blomberg; **Photographer (for cover and interior illustration references):** Neil Bernstein; **Visual Production Assistant:** Joyce Brumfield; **Art Manager:** Kelly Hendren; **Associate Art Manager:** Alan L. Wilborn; **Illustrator (cover):** Jen Gibas; **Illustrator (interior):** Molly Borman; **Printer:** Versa Press

Human Kinetics books are available at special discounts for bulk purchase. Special editions or book excerpts can also be created to specification. For details, contact the Special Sales Manager at Human Kinetics.

Printed in the United States of America   10 9 8 7 6 5

The paper in this book is certified under a sustainable forestry program.

Human Kinetics
Website: www.HumanKinetics.com

United States: Human Kinetics
P.O. Box 5076
Champaign, IL 61825-5076
800-747-4457
e-mail: info@hkusa.com

Canada: Human Kinetics
475 Devonshire Road Unit 100
Windsor, ON N8Y 2L5
800-465-7301 (in Canada only)
e-mail: info@hkcanada.com

Europe: Human Kinetics
107 Bradford Road
Stanningley
Leeds LS28 6AT, United Kingdom
+44 (0) 113 255 5665
e-mail: hk@hkeurope.com

Australia: Human Kinetics
57A Price Avenue
Lower Mitcham, South Australia 5062
08 8372 0999
e-mail: info@hkaustralia.com

New Zealand: Human Kinetics
P.O. Box 80
Mitcham Shopping Centre, South Australia 5062
0800 222 062
e-mail: info@hknewzealand.com

E5800

CONTENTS

INTRODUCTION

Flexibility is an important component of overall fitness. Unfortunately, flexibility is generally not one of the main focuses of many fitness programs. It is usually given very little attention or is neglected altogether. Although the benefits of regular exercise are well known, few people realize that flexible joints and regular stretching are also essential for optimal health and activity. For example, stretching can help people who have arthritis. To help relieve pain, especially during the early stages of this condition, people who have arthritis often keep affected joints bent and still. Although holding a joint still and bent may temporarily relieve discomfort, keeping a joint in the same position causes the muscles and ligaments to stiffen. This lack of movement can cause the muscles to shorten and become tight, leading to permanent loss of mobility and a hindering of daily activities. In addition, less movement means fewer calories burned, and any added weight puts more strain on the joints. Therefore, fitness experts urge people who have arthritis to stretch all of the major muscle groups daily, placing a gentle emphasis on joints that have decreased range of motion.

Good flexibility is known to bring positive benefits to the muscles and joints. It aids with injury prevention, helps minimize muscle soreness, and improves efficiency in all physical activities. This is especially true for people whose exercise sessions, whether a recreational game of golf or a more strenuous weekend game of basketball, are more than four days apart. Increasing flexibility can also improve quality of life and functional independence. People whose daily lifestyle consists of long sessions of inactivity such as sitting at a desk can experience a stiffening of the joints so that it is difficult to straighten out from that chronic position. Good flexibility helps prevent this by maintaining the elasticity of the muscles and providing a wider range of movements in the joints. It also provides fluidity and ease in body movements and everyday activities. A simple daily task such as bending over and tying your shoes is easier when you have good flexibility.

Stretching can also help prevent and relieve many muscle cramps, especially leg cramps that occur during the night. The causes of nighttime leg cramps are varied: too much exercise; muscle overuse; standing on a hard surface for a long time; flat feet; sitting for a long time; an awkward leg position during sleep; insufficient potassium, calcium, or other minerals; dehydration; certain medicines such as antipsychotics, birth control pills, diuretics, statins, and steroids; and diabetes or thyroid disease. Regardless of the cause, a more flexible muscle is less likely to cramp, and stretching helps to immediately reduce the cramp. Interestingly, current research shows that people who have type 2 diabetes or who are at high risk can help control blood glucose levels by doing 30 to 40 minutes of stretching. Thus, it is easy to see the benefits of making a stretching program a daily habit.

How much stretching should the average person do every day? Most people tend to overlook this important fitness routine altogether. Those who do stretch tend to perform a very brief routine that concentrates mainly on the lower-body muscle groups. In fact, it would be generous to suggest that people stretch any particular muscle group for more than 15 seconds. The total time spent in a stretching routine hardly ever exceeds 5 minutes. Even in athletics, stretching is given minor importance in the overall training program. An athlete might spend just a little more time stretching than the average person, usually because stretching is part of a warm-up routine. After the workout, however, most athletes are either too tired to do any stretching or simply do not take the time to do it. Stretching can be performed both as part of the warm-up before a workout and as part of a cool-down after, although stretching as part of a warm-up has become controversial. Stretching right before an event can have negative consequences on athletic performance. These negative consequences are most evident if the stretching exceeds 30 seconds. Therefore, a short stretch or quick loosen-up can be part of the warm-up, but stretching to induce permanent increases in flexibility should be done as part of the cool-down.

ANATOMY AND PHYSIOLOGY OF STRETCHING

Muscles such as the biceps brachii are complex organs composed of nerves, blood vessels, tendons, fascia, and muscle cells. Nerve cells (neurons) and muscle cells are electrically charged. The resting electrical charge, or resting membrane potential, is negative and is generally around –70 millivolts. Neurons and muscle cells are activated by changing their electrical charges. Electrical signals cannot jump between cells, so neurons communicate with other neurons and with muscle cells by releasing specialized chemicals called *neurotransmitters*. Neurotransmitters work by enabling positive sodium ions to enter the cells and make the resting membrane potential more positive. Once the resting membrane potential reaches a threshold potential (generally –62 millivolts), the cell becomes excited, or active. Activated neurons release other neurotransmitters to activate other nerves, causing activated muscle cells to contract.

Besides being altered to cause excitation, the membrane potential can be altered to cause either facilitation or inhibition. Facilitation occurs when the resting membrane potential is raised slightly above normal but below the threshold potential. Facilitation increases the likelihood that any succeeding neurotransmitter releases will cause the potential to exceed the threshold. This enhances the chances of the neuron's firing and activating the target. Inhibition occurs when the resting membrane potential is lowered below the normal potential, thereby decreasing the likelihood of reaching the threshold. Usually this prevents the neuron from activating its target.

To perform work, the muscle is subdivided into motor units. The motor unit is the basic functional unit of the muscle. A motor unit consists of one motor (muscle) neuron and all the muscle cells to which it connects, as few as 4 to more than 200. Motor units are then subdivided into individual muscle cells. A single muscle cell is sometimes referred to as a *fiber*. A muscle fiber is a bundle of rodlike structures called *myofibrils* that are surrounded by a network of tubes known as the sarcoplasmic reticulum, or SR. Myofibrils are formed by a series of repeating structures called *sarcomeres*. Sarcomeres are the basic functional contractile units of a muscle.

The three basic parts of a sarcomere are thick filaments, thin filaments, and Z-lines. A sarcomere is defined as the segment between two neighboring Z-lines. The thin filaments are attached to both sides of a Z-line and extend out from the Z-line for less than one-half of the total length of the sarcomere. The thick filaments are anchored in the middle of the sarcomere. Each end of a single thick filament is surrounded by six thin filaments in a helical array. During muscle work (concentric, eccentric, or isometric), the thick filaments control the amount and direction that the thin filaments slide over the thick filaments. In concentric work, the thin filaments slide toward each other. In eccentric work, the thick filaments try to prevent the thin filaments from sliding apart. For isometric work, the filaments do not move. All forms of work are initiated by the release of calcium ions from the SR, which occurs only when the muscle cell's resting membrane potential exceeds the threshold potential. The muscle relaxes and quits working when the calcium ions are restored within the SR.

The initial length of a sarcomere is an important factor in muscle function. The amount of force produced by each sarcomere is influenced by length in a pattern similar in shape to an upside-down letter U. As such, force is reduced when the sarcomere length is either long or short. As the sarcomere lengthens, only the tips of the thick and thin filaments can contact each other, and this reduces the number of force-producing connections between the two filaments. When the sarcomere shortens, the thin filaments start to overlap each other, and this overlap also reduces the number of positive force-producing connections.

Sarcomere length is controlled by proprioceptors, or specialized structures incorporated within the muscle organs, especially within the muscles of the limbs. The proprioceptors are specialized sensors that provide information about joint angle, muscle length, and muscle tension. Information about changes in muscle length is provided by proprioceptors called muscle spindles, and they lie parallel to the muscle cells. The Golgi tendon organs, or GTOs, the other type of proprioceptor, lie in series with the muscle cells. GTOs provide information about changes in muscle tension and indirectly can influence muscle length. The muscle spindle has a fast dynamic component and a slow static component that provides information on the amount and rate of change in length. Fast length changes can trigger a stretch, or myotatic, reflex that attempts to resist the change in muscle length by causing the stretched muscle to contract. Slower stretches allow the muscle spindles to relax and adapt to the new longer length.

When the muscle contracts it produces tension in the tendon and the GTOs. The GTOs record the change and rate of change in tension. When this tension exceeds a certain threshold, it triggers the lengthening reaction via spinal cord connections to inhibit the muscles from contracting and cause them to relax. Also, muscle contraction can induce reciprocal inhibition, or the relaxation of the opposing muscles. For instance, a hard contraction of the biceps brachii can induce relaxation within the triceps brachii.

The body adapts differently to acute stretching (or short-term stretching) and chronic stretching (or stretching done multiple times during a week). The majority of current research shows that when acute stretches cause a noticeable increase in a joint's range of motion, the person can experience either inhibition of the motor nerves, overlengthening of the muscle sarcomeres, or increased length and compliance of the muscle's tendons. No one is sure of the extent of these changes, but it appears that the muscle shape and cell arrangement, muscle length and contribution to movement, and length of the distal and proximal tendons all play a role. Nevertheless, these transient changes are manifested as decreases in maximal strength, power, and strength endurance. On the other hand, research studies have shown that regular heavy stretching for a minimum of 10 to 15 minutes three or four days a week (chronic stretching) results in the development of increased strength, power, and strength endurance as well as improved flexibility and mobility. Animal studies suggest that these benefits are due in part to increased numbers of sarcomeres in series.

Likewise, research into stretching for injury prevention has shown differences between acute stretching and chronic stretching. Although acute stretching can help an extremely tight person reduce the incidence of muscle strains, the normal person appears to gain minimal injury-prevention benefit from acute stretching. People who are inherently more flexible are less prone to exercise-related injuries, and inherent flexibility is increased with heavy stretching three or four days a week. Because of these differences between acute and chronic stretching, many exercise experts now encourage people to do the majority of their stretching at the end of a workout.

TYPES OF STRETCHES

The stretches featured in this book can be executed in a variety of ways. Most people prefer to do these stretches on their own, but they can also be done with the help of another person. Stretches performed without assistance are referred to as active stretches. Stretches performed with assistance from another person are called passive stretches.

There are four major types of stretches: static, ballistic, proprioceptive neuromuscular facilitation (PNF), and dynamic.

The *static stretch* is the most common. In static stretching, you stretch a particular muscle or group of muscles by holding that stretch for a period of time.

Ballistic stretches involve bouncing movements and do not involve holding the stretch for any length of time. Since ballistic stretching can activate the stretch reflex, many people have postulated that ballistic stretching has a greater potential to cause muscle or tendon damage, especially in the tightest muscles. However, this assertion is purely speculative, and no published research supports the claim that ballistic stretching can cause injury.

Proprioceptive neuromuscular facilitation (PNF) stretching refers to a stretching technique that tries to more fully incorporate the actions of the proprioceptors by stretching a contracted muscle through the joint's range of motion. After moving through the complete range of motion, the muscle is relaxed and rested before it is stretched again. This type of stretching is best done with the assistance of another person.

Dynamic stretching is a more functionally oriented stretch that uses sport-specific movements to move the limbs through a greater range of motion than normal. Dynamic stretching is generally characterized by swinging, jumping, or exaggerated movements in which the momentum of the movement carries the limbs to or past the regular limits of the range of motion and activates a proprioceptive reflex response. The proper activation of the proprioceptors can cause facilitation of the nerves that activated the muscle cells. This facilitation enables the nerves to fire more quickly, thus enabling the muscle to make fast and more powerful contractions. Since dynamic stretches increase both muscle temperature and proprioceptive activation, dynamic stretching has been found to be advantageous for improving athletic performance. Dynamic stretching should not be confused with ballistic stretching. Although both involve repeated movements, ballistic movements are rapid, bouncing movements that involve small ranges of movement near the end of the range of motion.

BENEFITS OF A STRETCHING PROGRAM

Several chronic training benefits can be gained through a regular stretching program (see chapter 9 for specific programs):

- Improved flexibility, stamina (muscular endurance), and muscular strength (the degree of benefit depends on how much stress is put on the muscle; chapter 9 explains how this should be done)
- Reduced muscle soreness
- Improved muscle and joint mobility
- More efficient muscular movements and fluidity of motion
- Greater ability to exert maximum force through a wider range of motion
- Prevention of some lower-back problems
- Improved appearance and self-image
- Improved body alignment and posture
- Better warm-up and cool-down in an exercise session
- Improved maintenance of blood glucose

STATIC AND DYNAMIC STRETCHES FOR ATHLETES

Many athletes use static and dynamic stretches in their training programs. Static stretches improve flexibility in certain muscle–joint areas. This type of stretching is the most common approach for improving flexibility. In static stretching, you hold a stretch of a particular muscle or muscle group for a period of time.

Some athletes prefer using dynamic stretches, particularly as a part of a warm-up or as a preparation for competition. Dynamic stretches stimulate the proprioceptors (stretch receptors), activating their response in an aggressive way by sending feedback to the stretched muscles to be contracted after a quick bouncing motion. Because some athletic events, such as explosive, short-duration activities, could possibly enhance the stimulation of this proprioceptive activation, dynamic stretching prepares athletes better for explosive movements. Such explosive movements might be required to accomplish a certain goal in an athletic event. For example, a person can jump farther and higher if he does a couple of quick up and down movements, flexing and extending the hips and knees.

HOW TO USE THIS BOOK

The first seven chapters of this book highlight stretches for the major joint areas of the body, beginning with the neck and ending with the feet and calves. Within each chapter are several stretches targeting the muscles involved in moving the joints in each part of the body. The movements targeting what are likely to be the stiffest muscles include a progression of stretches so that the person with the tightest muscles (beginner) is not trying to do a stretch that puts too much stress on the joint and results in muscle, ligament, and tendon damage. As you increase in flexibility, graduate to the next level.

Chapter 8 contains nine dynamic stretches that encompass all the major joint areas. Chapter 9 contains suggested stretching programs for beginners through advanced as well as a program shown to lower blood glucose. In addition, chapter 9 includes sport-specific stretching routines. If you are interested in a specific sport, these tables will guide you to the stretches to use in your training to ensure that you target the most important muscle groups used in that sport.

The name of each stretch indicates the major movements of the muscles being stretched. As such, you should remember that to stretch a specific muscle, the stretch must involve one or more movements in the opposite direction of the desired muscle's movements. The illustrations depict the body positions used for each stretch as well as the muscles being stretched. The muscles most stretched are illustrated in a dark red (see key), and any nearby muscles that are less stretched are illustrated in a lighter red.

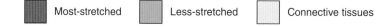

In addition to the illustrations, each stretch contains three sections:

- Execution, which provides step-by-step instructions on how to perform the stretch
- Muscles stretched, which provides the names of the muscles being stretched
- Stretch notes, which provide specific information concerning the how and why behind the need for the stretch as well as any safety considerations

chapter 1

NECK

The seven cervical vertebrae along with associated muscles and ligaments make up the flexible framework of the neck. The vertebrae, muscles, and ligaments work together to support and move the head. The first and second cervical vertebrae have unique shapes and are called the *atlas* and *axis*. The atlas is a bony ring that supports the skull. The axis has an upward peglike projection, the dens, that gives the atlas a point to pivot around. The axis and the other five cervical vertebrae have a posterior bony protuberance, or spinous process, that attaches to the large, thick nuchal ligament. The vertebral bodies (the oval-shaped bone mass) are connected by posterior and anterior ligaments, along with other ligaments that connect each spinous and transverse (lateral bony protuberance) process to their corresponding parts on the adjacent vertebrae. In addition, each vertebra is separated by an intervertebral disc. Through compression of the vertebrae upon the discs, the neck can move forward, backward, and sideways.

The neck muscles are located in two triangular regions called the anterior (front) and posterior (back) triangles. The borders of the anterior triangle are the mandible (jawbone), the sternum (breastbone), and the sternocleidomastoid muscle. The major anterior muscles are the sternocleidomastoid and scalene (figure 1.1*a*). The borders of the posterior triangle are the clavicle (collarbone), sternocleidomastoid muscle, and trapezius muscle. The major posterior muscles (figure 1.1*b*) are the trapezius, longissimus capitis, semispinalis capitis, and splenius capitis.

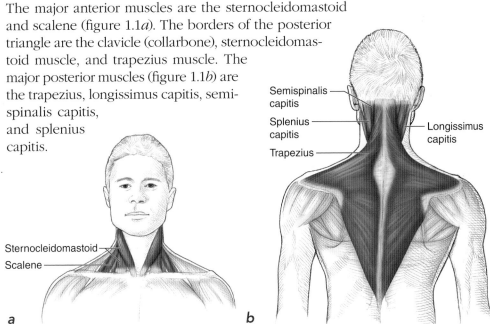

Semispinalis capitis

Splenius capitis

Trapezius

Longissimus capitis

Sternocleidomastoid

Scalene

a   *b*

Figure 1.1   Neck muscles: *(a)* anterior; *(b)* posterior.

The head movements are flexion (head tilted forward), extension (head tilted backward), lateral flexion and extension (head tipped from side to side), and rotation. Since the muscles in the neck come in right and left pairings, all the neck muscles are involved in lateral flexion and extension. For example, the right sternocleidomastoid helps perform right lateral flexion, and the left ster-nocleidomastoid helps perform right lateral extension. Neck flexion is limited not only by the stiffness of the posterior muscles but also by the stiffness of the posterior ligaments, the strength of the flexor muscles, the alignment of the vertebral bodies with the adjacent vertebrae, the compressibility of the anterior portions of the intervertebral discs, and the contact of the chin with the chest. Similarly, neck extension is controlled by the stiffness of the anterior muscles as well as by the stiffness of the anterior ligaments, the strength of the extensor muscles, the alignment of the vertebral bodies with the adjacent vertebrae, and the compressibility of the posterior portions of the intervertebral discs. Finally, in addition to the stiffness of the contralateral muscles and tendons, neck lateral function is controlled by the impingement of each vertebra's transverse process upon the adjacent transverse process.

People seldom consider the neck muscles when stretching. Neck flexibility probably does not cross your mind until you discover that you have a stiff neck. A stiff neck is commonly associated with sleeping in a strange position (such as on a long flight) or sitting at a desk for an extended time, but a stiff neck can result from almost any type of physical activity. This is especially true for any activity in which the head must be held in a constantly stable position. A stiff neck can also have a negative effect in sports in which head position is impor-tant, such as golf, or when rapid head movements are important for tracking the flight of an object, such as in racket sports. Poor neck flexibility usually results from holding the head in the same position for long periods. In addition, a fatigued neck muscle can stiffen up after exercise. The exercises in this chapter can help keep the neck from stiffening up after exercises, unusual postures, or awkward sleep positions.

Since all the major muscles in the neck are involved in neck rotation, it is fairly easy to stretch the neck muscles. The first consideration when choosing a particular neck stretch should be whether greater stiffness occurs with flexion or extension. Therefore, the first two exercise groups focus on these specific actions. Once you achieve greater flexibility in either pure flexion or pure extension, then you can add a stretch that includes lateral movement. In other words, to increase the flexibility of the neck extensors, start with the neck extensor stretch and then, as flexibility increases, add the neck extensor and rotation stretch.

Stretching the neck can be dangerous if not done properly. Some stretches of the neck use what is termed a plow position in which the back of the head lies on a surface, with the trunk nearly perpendicular. This position can generate high stress at the bending point, especially in people with low neck flexibility. This high stress can either damage the vertebrae or greatly compress the anterior intervertebral disc. Disc compression can cause protrusion and pressure on the spinal cord, thus damaging it. Additionally, when stretching the neck, a person must be careful not to apply sudden or rapid force. Sudden force application can lead to whiplash injuries; in the worst-case scenario, whiplash can sever the vertebral arteries and force the dens into the brain's medulla oblongata, causing death.

Also, be aware that overstretching or doing very hard stretching causes more harm than good. Sometimes a muscle becomes stiff from overstretching. Stretching can reduce muscle tone, and when tone is reduced, the body compensates by making the muscle even tighter. For each progression, start with the position that is the least stiff and progress only when, after several days of stretching, you notice a consistent lack of stiffness during the exercise. This means you should stretch both the agonist muscles (the muscles that cause a movement) and antagonist muscles (the muscles that oppose a movement or do the opposite movement). And although you may have greater stiffness in one direction (right versus left), you need to stretch both sides so that you maintain proper muscle balance.

The stretches in this chapter are excellent overall stretches; however, not all of these stretches may be completely suited to each person's needs. To stretch specific muscles, the stretch must involve one or more movements in the opposite direction of the desired muscle's movements. For example, if you want to stretch the left scalene, you could extend the head both back and laterally to the right. When a muscle has a high level of stiffness, you should use fewer simultaneous opposite movements. For example, you would stretch a very tight right scalene by initially doing just left lateral extension. As a muscle becomes loose, you can incorporate more simultaneous opposite movements.

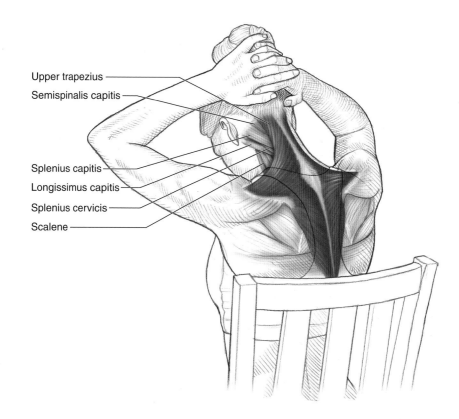

Upper trapezius
Semispinalis capitis

Splenius capitis
Longissimus capitis
Splenius cervicis
Scalene

Execution

1. Sit comfortably with the back straight.
2. Interlock the hands on the back of the head near the crown.
3. Lightly pull the head straight down and try to touch the chin to the chest.

Muscles Stretched

Most-stretched muscle: Upper trapezius

Less-stretched muscles: Longissimus capitis, semispinalis capitis, splenius capitis, splenius cervicis, scalene

Stretch Notes

You can do this stretch while either sitting or standing. A greater stretch is applied when seated. Standing reduces the ability to stretch because reflexes come into play to prevent a loss of balance. Therefore we recommend doing the stretch while seated. During the stretch, make sure not to reduce the stretch by hunching up the shoulders. Also, keep the neck as straight as possible (no curving). Try to touch the chin to the lowest possible point on the chest.

It is common for people who are stressed to hunch their shoulders. Constantly hunching does not allow the posterior neck muscles any chance to relax. This causes these muscles to become tight, adding to the pain and fatigue and causing more hunching. Additionally, these muscles can become tight after any neck strain or whiplash injury. Relief and relaxation can be obtained by doing this stretch, thus greatly decreasing hunching. Also, the neck extensor muscles must remain loose in order to maintain proper posture, and maintaining proper posture can in turn help reduce muscle strain and tightness.

NECK EXTENSOR AND ROTATION STRETCH

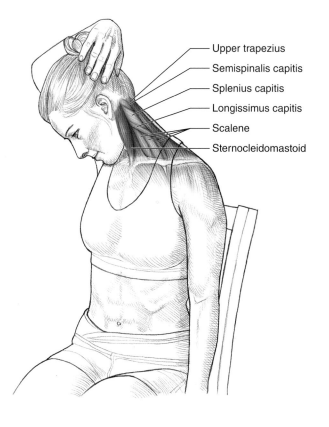

Upper trapezius
Semispinalis capitis
Splenius capitis
Longissimus capitis
Scalene
Sternocleidomastoid

Execution

1. Sit comfortably with the back straight.
2. Place the right hand on the back of the head near the crown.
3. Pull the head down and to the right so that it points to the right shoulder. Bring the chin as close to the right shoulder as possible.
4. Repeat the stretch on the other side.

Muscles Stretched

Most-stretched muscles: Left upper trapezius, left sternocleidomastoid

Less-stretched muscles: Left longissimus capitis, left semispinalis capitis, left splenius capitis, left scalene

Stretch Notes

After the neck extensors become flexible, you can progress from stretching both sides of the neck simultaneously to stretching the left and right sides individually. Stretching one side at a time allows you to place a greater stretch on the muscles. Often one side of the neck is stiffer than the other side. Frequently this occurs if you sleep strictly on one side or sit at a desk and do not look straight ahead but continually look either to the left or the right.

When you stretch both sides of the neck simultaneously, the amount of stretch applied is limited by the stiffest muscles. Thus, if one side is more flexible, it may not receive a sufficient stretch. By stretching each side individually, you can concentrate more effort on the stiffer side.

You can perform this stretch while either sitting or standing. Although you can achieve a better stretch while sitting, do whichever feels best to you.

NECK FLEXOR STRETCH

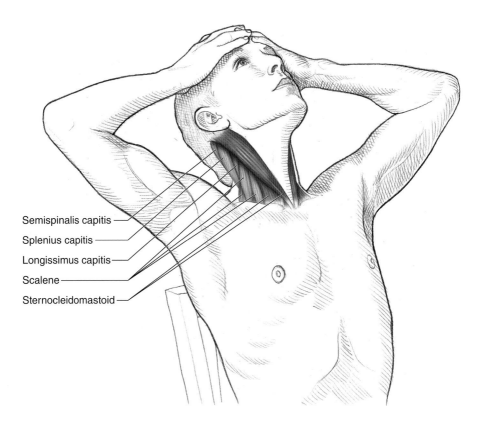

Semispinalis capitis
Splenius capitis
Longissimus capitis
Scalene
Sternocleidomastoid

Execution

1. Sit comfortably with the back straight.
2. Interlock the hands and place the palms on the forehead.
3. Pull the head back so that the nose points straight up to the ceiling.

Muscles Stretched

Most-stretched muscle: Sternocleidomastoid

Less-stretched muscles: Longissimus capitis, semispinalis capitis, splenius capitis, scalene

Stretch Notes

You can do this stretch while either sitting or standing. A greater stretch is applied when seated. Standing reduces the ability to stretch because reflexes come into play to prevent a loss of balance. Therefore we recommend doing the stretch while seated. During the stretch, make sure not to reduce the stretch by hunching the shoulders. Also try to bring the chin as far back as possible.

When people are under stress, they typically breathe forcefully while keeping their shoulders raised. This can lead to pain and tension in the anterior neck muscles. Short-term relief can be obtained by doing this stretch. Also, the neck flexor muscles must remain loose in order to maintain proper posture. If you let these muscles become tight, you can end up with the deformation commonly called vulture neck, in which the head position looks like the protruding head of a vulture. To help maintain correct posture, this stretch should be done several times a week.

NECK FLEXOR AND ROTATION STRETCH

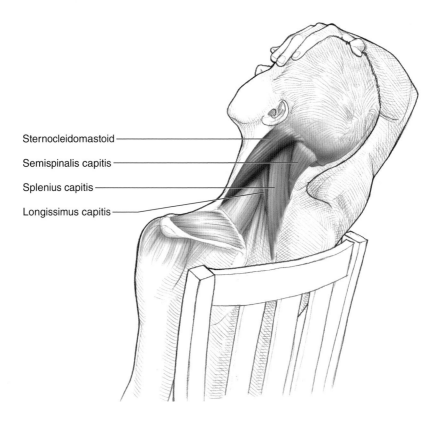

Sternocleidomastoid

Semispinalis capitis

Splenius capitis

Longissimus capitis

Execution

1. Sit comfortably with the back straight.
2. Place the right hand on the forehead.
3. Pull the head back and toward the right so that the head points toward the shoulder.
4. Repeat for the left side.

Muscles Stretched

Most-stretched muscle: Left sternocleidomastoid

Less-stretched muscles: Left longissimus capitis, left semispinalis capitis, left splenius capitis

Stretch Notes

After the neck flexors become flexible, progress from stretching both sides of the neck simultaneously to stretching the left and right sides individually. Stretching one side at a time allows you to place a greater stretch on the muscles. This especially is important for those who stand hunched over with the head pointed mainly to one side.

When you stretch both sides of the neck simultaneously, the amount of stretch applied is limited by the stiffest muscles. Thus, the more flexible side may not receive a sufficient stretch. By stretching each side individually, you can concentrate more effort on the stiffer side.

You can perform this stretch while either sitting or standing upright. Although you can achieve a better stretch while sitting, choose whichever position feels best to you.

chapter 2

SHOULDERS, BACK, AND CHEST

There are five major pairs of movements at the shoulder: (1) flexion and extension, (2) abduction and adduction, (3) external and internal rotation, (4) retraction and protraction, and (5) elevation and depression. The bones of the shoulder joint consist of the humerus (upper-arm bone), scapula (shoulder blade), and clavicle (collarbone). The scapula and clavicle essentially float on top of the rib cage. Therefore, a major function of many upper-back and chest muscles is to attach the scapula in the upper back and the clavicle in the upper chest to the rib cage and spine. This provides a stable platform for arm and shoulder movements. Of the five movement pairs, retraction and protraction and elevation and depression usually are classified as stabilization actions.

Most of the muscles involved in moving and stabilizing the shoulder bones are located posteriorly. The scapula is a much larger bone than the clavicle and has room for more muscles to attach. The posterior (back) muscles (figure 2.1) are the infraspinatus, latissimus dorsi, levator scapulae, rhomboids, subscapularis, supraspinatus, teres major, teres minor, and trapezius (attached to the upper posterior rib cage, vertebrae, and scapula), as well as the deltoid (figure 2.2) and triceps brachii (attached to the scapula and humerus; see chapter 3). The anterior (front) muscles (figure 2.3) are the pectoralis major (attached to the clavicle, anterior rib cage, and humerus), pectoralis minor, subclavius, serratus anterior (attached to the anterior rib cage and anterior scapula), biceps brachii, coracobrachialis, and deltoid (attached to the anterior scapula and humerus).

The shoulder, or glenohumeral, joint is a ball-and-socket joint formed by the head of the humerus and the glenoid fossa, a shallow scapular cavity that forms a socket for the humeral head. This joint is both the most freely moving joint of the body and the least stable. Upward movement of the humerus is prevented by the clavicle and the scapular acromion and coracoid processes, as well as by the glenohumeral ligaments and rotator cuff. Downward, forward, and backward humeral movements are limited by the humeral head's position in the glenoid labrum, a circular band of fibrocartilage that passes around the rim of the glenoid fossa to increase its concavity. Along with the glenoid labrum, the humerus is held in place by several ligaments and muscle tendons that together form the rotator cuff.

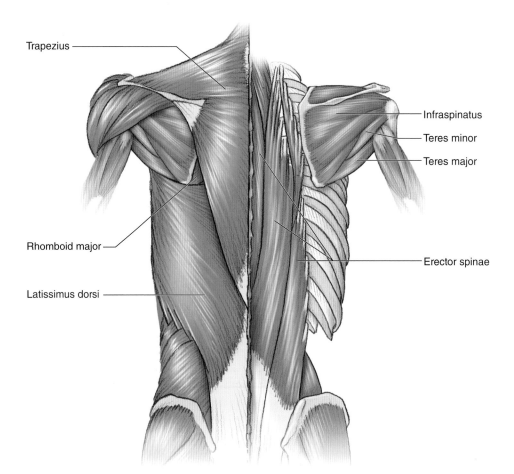

Trapezius

Infraspinatus

Teres minor

Teres major

Rhomboid major

Erector spinae

Latissimus dorsi

Figure 2.1 Back muscles.

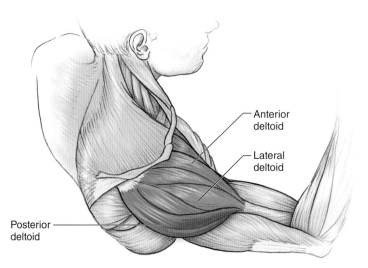

Anterior deltoid

Lateral deltoid

Posterior deltoid

Figure 2.2 Deltoid muscle.

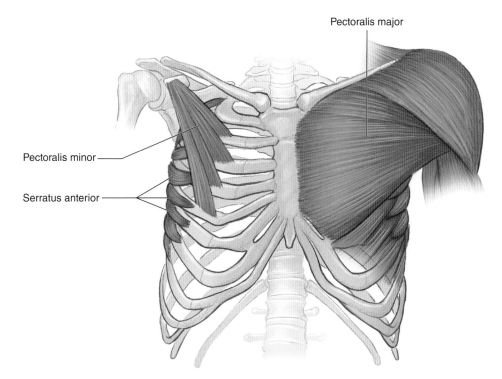

Figure 2.3 Chest muscles.

The whole humerus head and the glenoid fossa are surrounded by the joint capsule, a collection of ligaments. Major ligaments include the anterior and posterior sternoclavicular, costoclavicular, and interclavicular ligaments, which help connect the clavicle to the rib cage. The coracohumeral, glenohumeral, coracoclavicular, acromioclavicular, and coracoacromial ligaments help interconnect the humerus, scapula, and clavicle bones. The major muscles and tendons providing rotator cuff stability are the infraspinatus, subscapularis, supraspinatus, and teres minor. Since these muscles attach more superiorly (atop the shoulder), most dislocations occur inferiorly (downward from the shoulder).

Since the shoulder muscles are a major component of shoulder stability, shoulder flexibility—the amount of possible movement in a particular direction—in all five movement pairs (e.g., extension and flexion) is greatly controlled by both the strength of the muscles and the extensibility of the antagonist muscles involved in the movement. Shoulder abduction, the range of motion away from the midline of the body, is limited by the flexibility of the ligaments in both the shoulder and the joint capsule and by the humerus hitting the acromion and the superior rim of the glenoid fossa (or shoulder impingement). Shoulder adduction, the range of motion toward the midline of the body, is additionally limited by the arm meeting the trunk. Shoulder flexion range of motion is limited by the tightness of both the coracohumeral ligament and the inferior portion of the joint capsule. Coracohumeral ligament flexibility influences shoulder extension range of motion along with shoulder impingement. Shoulder internal rotation is restricted by the flexibility of the capsular ligaments, while external rotation range of movement is limited by rigidity of the coracohumeral ligament and the tightness of the superior portion of the capsular ligaments. Additional factors for elevation include the tension of the costoclavicular ligament along with the joint capsule. For depression the other restrictors are the interclavicular and sternoclavicular ligaments. Finally, protraction is limited by tightness in both the anterior sternoclavicular and posterior costoclavicular ligaments, while retraction is limited by tightness in both the posterior sternoclavicular and anterior costoclavicular ligaments.

It is important to maintain proper balance between strength and flexibility in all shoulder muscles. Common complaints associated with the musculature of the shoulders, back, and chest are tight muscles and muscle spasms in the neck (middle and upper trapezius), shoulder (trapezius, deltoid, supraspinatus), and upper back (rhomboids and levator scapulae). Interestingly, the tightness felt in these muscles is usually a result of initial tightness in their antagonist muscles. In other words, tight muscles in the upper chest caused the tightness felt in the upper back. Tight chest muscles (e.g., the pectoralis major) cause a constant low-level stretch on the muscles of the upper back. Eventually, this low-level stretch elongates the ligaments and tendons associated with the upper-back muscles. Once these ligaments and tendons become elongated, the tone in their associated muscles falls dramatically. To reclaim the lost tone, the muscles must increase their force of contraction. Increased force in turn causes more stretch of the ligaments and tendons, and increased muscle contraction must compensate for that. Hence, a vicious cycle commences.

The best way to prevent or stop this cycle is to stretch the anterior shoulder and chest muscles. As the flexibility of these muscles increases, the tightness of the posterior muscles is reduced. Immediately after stretching, the strength of the muscles is diminished. It is a good idea to stretch the opposing muscles just before and immediately after working any group of muscles. If this is done three or more times a week, the muscles will actually increase in flexibility and gain strength. Stretching will also reduce the frequency of tightness for any group of muscles. Furthermore, shoulder impingement can occur with improper balance between shoulder muscle strength and flexibility. Since the gap between the humerus and scapular process is narrow, anything that further narrows this space, such as tight muscles, can result in impingement, leading to pain, weakness, and loss of movement.

Many of the instructions and illustrations in this chapter are given for the left side of the body. Similar but opposite procedures would be used for the right side of the body. Although the stretches in this chapter are excellent overall stretches, some people may need additional stretches. Remember to stretch specific muscles, and the stretch must involve one or more movements in the opposite direction of the desired muscle's movements. For example, if you want to stretch the serratus anterior, perform a movement that involves shoulder depression, shoulder retraction, and shoulder adduction. When any muscle has a high level of stiffness, you should use very few simultaneous opposite movements. For example, to stretch a very tight pectoralis major, start by doing shoulder extension and external rotation. As a muscle becomes loose, you can incorporate more simultaneous opposite movements.

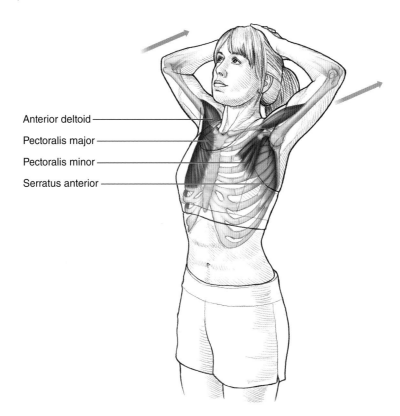

Anterior deltoid

Pectoralis major

Pectoralis minor

Serratus anterior

Execution

1. Stand upright and interlock your fingers.
2. Place your hands on top of your head.
3. Contract your back muscles, and pull your elbows back toward each other.

Muscles Stretched

Most-stretched muscles: Pectoralis major, pectoralis minor, anterior deltoid

Less-stretched muscle: Serratus anterior

Stretch Notes

Poor posture is the primary reason for tight shoulder flexor muscles. Poor posture is commonly seen when the person hunches forward or works with his arms extended out in front. Tightness usually is accompanied by tight neck extensors. Having both groups of muscles tight increases the chances of developing a vulture neck and contributes to breathing problems. Injuries, either acute or overuse, that lead to shoulder impingement, shoulder bursitis, rotator cuff tendinitis, or frozen shoulder can also lead to tight shoulder flexors.

When any of these conditions are severe, it is difficult to stretch the flexors without pain. This stretching activity places a low stretch stress on the musculature and hence is easy to tolerate. When you feel less stretch when doing this activity, it is best to advance to one of the other shoulder flexor stretching activities.

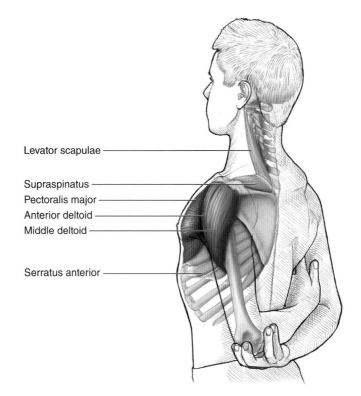

Levator scapulae

Supraspinatus
Pectoralis major
Anterior deltoid
Middle deltoid

Serratus anterior

Execution

1. Stand or sit upright on a backless chair, with the left arm behind the back and the elbow bent at about 90 degrees.
2. Place feet shoulder-width apart with the toes pointing forward.
3. Grasp the left elbow, forearm, or wrist, depending on your flexibility, with the right hand.
4. Pull the upper left arm across the back and up toward the right shoulder.
5. Repeat this stretch for the opposite arm.

Muscles Stretched

Most-stretched muscles: Left pectoralis major, left anterior deltoid, middle deltoid

Less-stretched muscles: Left levator scapulae, left pectoralis minor, left supraspinatus, left serratus anterior, left coracobrachialis

Stretch Notes

This stretch is excellent for overcoming a vulture neck or rounded, hunched shoulders arising from poor posture. It also helps relieve the pain associated with shoulder impingement, shoulder bursitis, rotator cuff tendinitis, and frozen shoulder. This exercise provides a better stretch than the beginner shoulder flexor stretch, but it is best to start using this stretch only after you have progressed through the beginner exercise and find it difficult to apply any stretch at the beginner level.

If you cannot reach the elbow, then grasp the wrist. When pulling on the wrist, it is easy to pull the arm across the back, but remember that the best effect comes from pulling upward as well as across. Also, keep the elbow locked at a near 90-degree angle. Changing the alignment of the back will also influence the magnitude of the stretch. If you cannot keep the back straight, arching the back is preferable to bending at the waist. Just be careful; it is easy to lose balance when doing this stretch while both arching the back and standing. If maintaining balance while standing is a problem, do this stretch while sitting on a stool or chair.

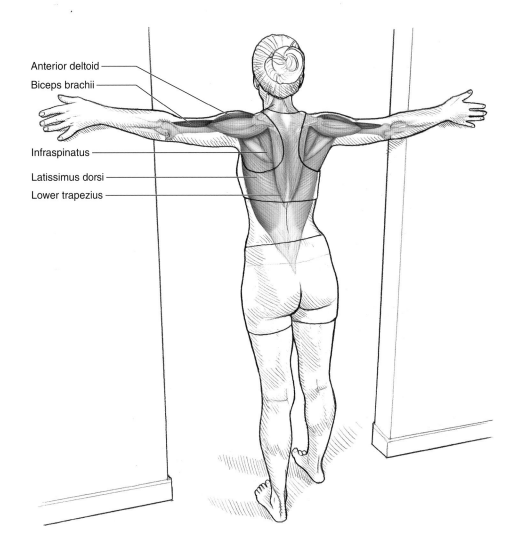

Anterior deltoid
Biceps brachii
Infraspinatus
Latissimus dorsi
Lower trapezius

Execution

1. Stand upright while facing a doorway or corner.
2. Place feet shoulder-width apart, with one foot slightly in front of the other.
3. With straight arms, raise your arms to shoulder level, and place the palms on the walls or doorframe with the thumbs on top.
4. Lean the entire body forward.

Muscles Stretched

Most-stretched muscles: Pectoralis major, anterior deltoid, coracobrachialis, biceps brachii

Less-stretched muscles: Infraspinatus, latissimus dorsi, subclavius, lower trapezius

Stretch Notes

This stretch is excellent for overcoming a vulture neck or rounded, hunched shoulders arising from poor posture. It also helps relieve the pain associated with shoulder impingement, shoulder bursitis, rotator cuff tendinitis, and frozen shoulder. However, if you have any of the aforementioned problems, it is better to start with the beginner stretch and work your way up to the advanced stretch. This exercise provides a better stretch than either the beginner or intermediate shoulder flexor stretches, and it is better to use if you can tolerate the pain or discomfort it may produce.

To get the maximum benefit during the stretch, keep the elbows locked and the spine straight. The greater the forward lean, the better the stretch. Forward lean is controlled by how far the lead foot is in front of the chest at the starting position. Hence, place the foot forward only enough to maintain balance. It is possible to do the neck extensor stretch simultaneously with the shoulder flexor stretch, but without the hands pushing down on the head. However, without having the hands pushing down on the head, the neck extensor stretch will be of a lower intensity than if it were done by itself.

⟨ **VARIATION** ⟩

Shoulder Flexor and Depressor Stretch

By elevating the arms above parallel, you can include the pectoralis minor as one of the major muscles being stretched. Stand upright while facing a doorway or corner, with the feet shoulder-width apart and one foot slightly in front of the other. Keeping the arms straight, raise the arms high above the head, and place the palms on the walls or doorframe. Lean the entire body forward.

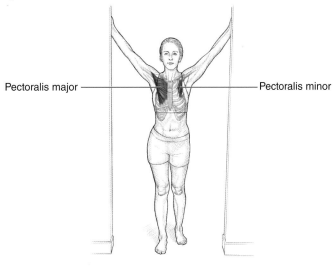

Pectoralis major — — Pectoralis minor

ASSISTED SHOULDER AND ELBOW FLEXOR STRETCH

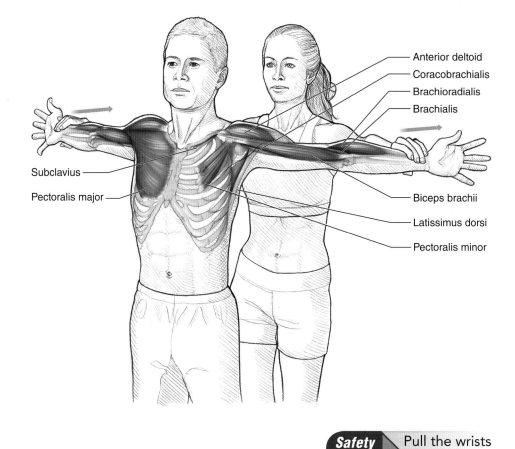

Anterior deltoid
Coracobrachialis
Brachioradialis
Brachialis

Subclavius

Pectoralis major

Biceps brachii

Latissimus dorsi

Pectoralis minor

Safety Tip Pull the wrists back gently.

Execution

1. Stand upright or sit on the floor for more stability.
2. If standing, place feet shoulder-width apart with one foot slightly in front of the other. If sitting, sit on the ground with both legs extended out in front of you.
3. Extend both arms parallel to the floor.
4. Point the hands slightly back.
5. Have a partner stand behind you facing your back and grab hold of each arm at the wrist.
6. The partner pulls the wrists toward each other while being careful not to overstretch the joint.

Muscles Stretched

Most-stretched muscles: Pectoralis major, pectoralis minor, anterior deltoid, coracobrachialis, biceps brachii, brachialis, brachioradialis

Less-stretched muscles: Latissimus dorsi, lower trapezius, subclavius

Stretch Notes

This stretch is excellent for overcoming a vulture neck or rounded, hunched shoulders arising from poor posture. It also helps relieve the pain associated with shoulder impingement, shoulder bursitis, rotator cuff tendinitis, and frozen shoulder. Additionally, this stretch helps prevent what many people call muscle boundness, or rounded and forward-thrusted shoulders combined with an inability to completely straighten the arms. This stretching activity is one of the better exercises for both the shoulder and elbow flexors. The partner can modify the stretch to tailor it to beginner through advanced by simply stretching to the point of pain toleration.

It is important for the partner assisting with this stretch not to become overly aggressive when pulling the wrists together. An overly aggressive stretch can result in muscle strains and, in extreme cases, shoulder dislocation. Moreover, as the wrists get closer to each other, people have a tendency to lean back to reduce the pain. If you find yourself leaning back, it is a good idea to bend at the waist and lean slightly forward at the start of the stretch.

SEATED SHOULDER FLEXOR, DEPRESSOR, AND RETRACTOR STRETCH

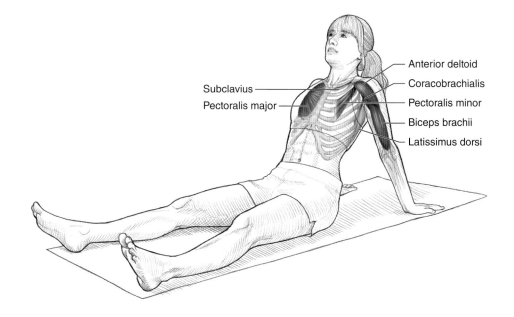

Subclavius
Pectoralis major
Anterior deltoid
Coracobrachialis
Pectoralis minor
Biceps brachii
Latissimus dorsi

Execution

1. Sit on the floor with the legs straight.
2. While keeping the arms straight, place the palms on the floor, fingers pointed back, about one foot (30 cm) behind the hips.
3. While keeping the arms straight, lean back toward the floor.

Muscles Stretched

Most-stretched muscles: Pectoralis major, anterior deltoid, coracobrachialis, biceps brachii, pectoralis minor

Less-stretched muscles: Latissimus dorsi, lower trapezius, subclavius, rhomboids

Stretch Notes

This stretching activity is one of the better unassisted exercises for stretching both the shoulder and elbow flexors simultaneously. It is an excellent stretch for overcoming a vulture neck or rounded, hunched shoulders arising from poor posture. It also helps relieve the pain associated with shoulder impingement, shoulder bursitis, rotator cuff tendinitis, and frozen shoulder. Additionally, this stretch helps prevent what many people call muscle boundness, or rounded and forward-thrusted shoulders combined with an inability to completely straighten the arms.

To maximize the stretch, keep the arms straight. If it is difficult to refrain from bending the arms, place the hands closer to the hips. Moving the hands farther from the hips can increase the stretch. To keep the body from sliding along the floor, you may need to brace the soles of the feet against a wall. Sitting on a mat with the hands on a hard surface will increase the stretch as well as add comfort.

BEGINNER SHOULDER EXTENSOR, ADDUCTOR, AND RETRACTOR STRETCH

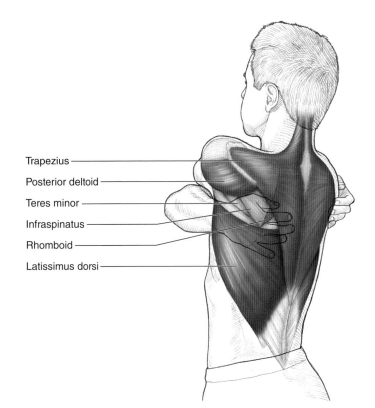

Trapezius
Posterior deltoid
Teres minor
Infraspinatus
Rhomboid
Latissimus dorsi

Execution

1. Stand upright with feet shoulder-width apart, toes pointing straight forward.
2. Wrap your arms around your shoulders as if you were hugging yourself, placing the arm on top that is the most comfortable.
3. Pull your shoulders forward.

Muscles Stretched

Most-stretched muscles: Posterior deltoid, latissimus dorsi, trapezius, rhomboids

Less-stretched muscles: Teres minor, infraspinatus

Stretch Notes

Poor posture overworks the deltoids, lats, traps, and rhomboids, causing tightness. This stretch relieves many of the aches and pains felt between the shoulder blades. Conversely, these muscles can also become tight from disuse or by doing limited activities with the arms below shoulder level. Tightness in these muscles makes any overhead work, such as painting a ceiling, washing overhead windows, or doing a dumbbell overhead press, harder and more painful. This stretching activity places a low stretch on the musculature and so is the best one to start with if you have extremely tight muscles. Also, doing this stretch helps relieve the pain associated with shoulder impingement, shoulder bursitis, rotator cuff tendinitis, and frozen shoulder.

INTERMEDIATE SHOULDER EXTENSOR, ADDUCTOR, AND RETRACTOR STRETCH

Middle deltoid
Posterior deltoid
Triceps brachii
Teres minor
Teres major
Latissimus dorsi
Serratus anterior

Execution

1. Stand upright inside a doorway while facing a doorjamb, with the door-jamb in line with the right shoulder.
2. Place feet shoulder-width apart, with the toes pointing straight forward.
3. Bring the left arm across the body toward the right shoulder.
4. Pointing the thumb down, grab hold of the doorjamb at shoulder level.
5. Rotate the trunk in until you feel a stretch in the posterior left shoulder.
6. Repeat these steps for the opposite arm.

Muscles Stretched

Most-stretched muscles: Left posterior deltoid, left middle deltoid, left latissimus dorsi, left triceps brachii, left middle trapezius, left rhomboids

Less-stretched muscles: Left teres major, left teres minor, left supraspinatus, left serratus anterior

Stretch Notes

Poor posture overworks the deltoids, lats, triceps, traps, and rhomboids, causing tightness. This intermediate stretch places more stretch on these muscles. It relieves many of the aches and pains felt between the shoulder blades better than the beginner stretch. Conversely, these muscles can also become tight from disuse or by doing limited activities with the arms below shoulder level. Tightness in these muscles makes any overhead work harder and more painful. This stretching activity places a greater stretch on the musculature than the basic shoulder extensor, adductor, and retractor stretch. Also, doing this stretch helps relieve the pain associated with shoulder impingement, shoulder bursitis, rotator cuff tendinitis, and frozen shoulder.

To get the maximum benefit of this stretch, you should keep the elbow locked. Over time, as the muscles become more flexible, to keep the elbow locked you will need to grasp the doorframe above the level of the shoulder. Raising the hand does not diminish the major benefits of this stretch. However, as the hand gets higher above shoulder level, the stretch on the rhomboids decreases while the stretch on the serratus anterior increases.

SHOULDER ADDUCTOR, PROTRACTOR, AND ELEVATOR STRETCH

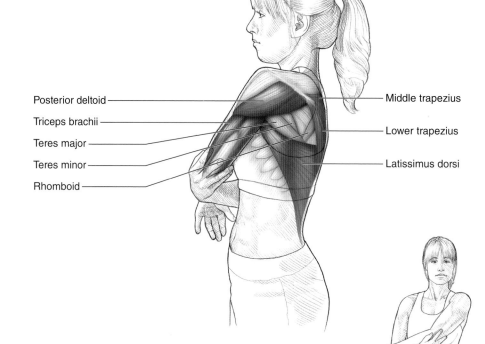

Posterior deltoid

Triceps brachii

Teres major

Teres minor

Rhomboid

Middle trapezius

Lower trapezius

Latissimus dorsi

Execution

1. Stand upright with the feet shoulder-width apart.
2. Bring the left arm across the front of the body, with the left hand near the right hip.
3. With the right hand, grab the left elbow.
4. With the right hand, try to pull the left elbow down and around the right side of the body.
5. Repeat these steps for the opposite arm.

Muscles Stretched

Most-stretched muscles: Left posterior deltoid, left latissimus dorsi, left triceps brachii, left lower middle trapezius

Less-stretched muscles: Left teres major, left teres minor, left supraspinatus, left levator scapulae, left rhomboids

Stretch Notes

Tightness in the deltoids, lats, triceps, and traps makes any overhead work harder and more painful. Thus this stretch makes it easier to do any throwing action as well as around-the-house activities such as painting and window cleaning. Also, doing this stretch can help relieve the pain associated with shoulder impingement, shoulder bursitis, rotator cuff tendinitis, and frozen shoulder.

To maximize the stretch, do not raise the shoulder or bend at the waist. If it is not possible to bring the hand toward the hip, try to come as close as possible. As long as the arm is below the shoulders, the stretch will be effective.

⟨ **VARIATION** ⟩

Overhead Shoulder Adductor, Protractor, and Elevator Stretch

Bringing the arm above the shoulder places more stretch on the elevators and protractors and is more beneficial for high overhead activities. Stand upright with the feet shoulder-width apart. Raise the left hand high above the head, and bring the left arm up against the left side of the head. Then, with the right hand, grab hold of the left elbow and try to pull the left elbow behind the head, past the left ear. Repeat these steps for the opposite arm.

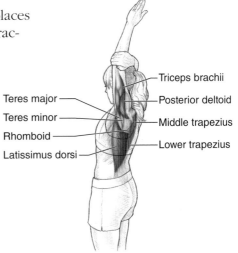

Teres major
Teres minor
Rhomboid
Latissimus dorsi
Triceps brachii
Posterior deltoid
Middle trapezius
Lower trapezius

SHOULDER ADDUCTOR AND EXTENSOR STRETCH

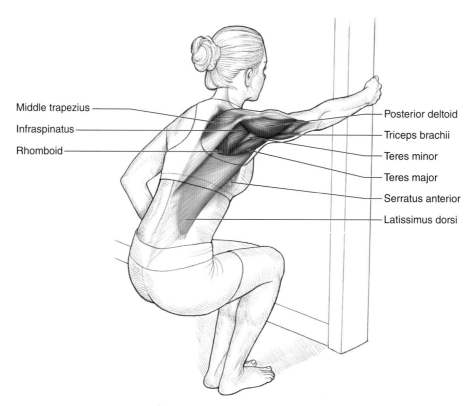

Middle trapezius

Infraspinatus

Rhomboid

Posterior deltoid

Triceps brachii

Teres minor

Teres major

Serratus anterior

Latissimus dorsi

Execution

1. Squat while facing a doorway, with the right shoulder lined up with the left side of the doorjamb.
2. Stick the right arm through the doorway. Grab the inside of the doorjamb at shoulder level with the right hand.
3. While keeping the right arm straight and the feet firmly planted, lower the buttocks toward the floor.
4. Repeat these steps for the opposite arm.

Muscles Stretched

Most-stretched muscles: Right posterior deltoid, right middle trapezius, right triceps brachii, right teres major, right rhomboids, right infraspinatus

Less-stretched muscles: Right latissimus dorsi, right teres minor, right supraspinatus, right serratus anterior

Stretch Notes

Although poor posture has a negative effect on both sides of the body and results in overall tightness, most people use one arm more than the other, so the muscles on one side can become tighter from disuse. This is especially possible when doing any overhead work such as painting, window washing, or overhead presses. These activities may become harder and more painful. Thus sometimes you may need to stretch one side more than the other. Since this stretch mimics single-arm overhead work, it is better suited for problems arising from one side being tighter than the other. Also, by stretching one side singularly with gravity assistance, this stretch allows for a greater amount of stretch than any of the other stretches that work similar muscles. Moreover, this stretch relieves many of the aches and pains felt between the shoulder blades.

A lower squat yields a greater stretch, but it increases the pressure and strain on the knee joints. Therefore, be careful not to squat so low that you feel pain in the legs or knees. To reduce strain on the knees, change the point where you grab the doorjamb. Changing the position of the grasp, however, influences the amount of stretch placed on the various muscles (see variation). Regardless of where you grasp, keep the back straight or arched. Do not bend forward at the waist. To get an even greater stretch, inwardly rotate the trunk.

⟨ VARIATION ⟩

Overhead Shoulder Adductor and Extensor Stretch

Grasping the inside of the doorjamb above head level reduces the stretch on the middle trapezius and allows a greater stretch of the posterior deltoid, latissimus dorsi, triceps brachii, teres major, and infraspinatus. Begin the stretch by squatting in front of a doorway, with the right shoulder in line with the left side of the doorjamb. Stick the right arm through the doorway, and, with the right hand, grab the inside of the doorjamb several inches above your head. Increase the stretch by lowering the buttocks toward the floor. Repeat again for the opposite side.

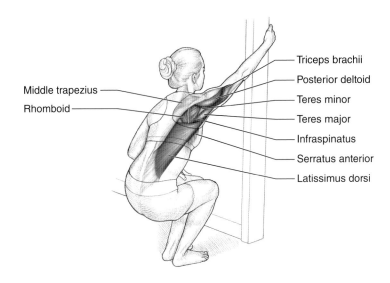

Middle trapezius

Rhomboid

Triceps brachii

Posterior deltoid

Teres minor

Teres major

Infraspinatus

Serratus anterior

Latissimus dorsi

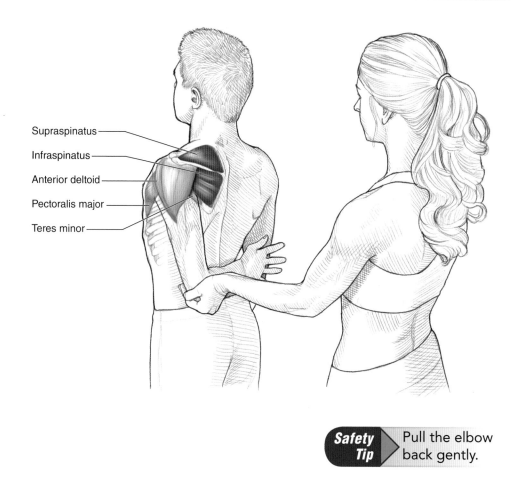

Supraspinatus

Infraspinatus

Anterior deltoid

Pectoralis major

Teres minor

Safety Tip Pull the elbow back gently.

Execution

1. Stand upright with feet shoulder-width apart, with the toes pointing straight forward.

2. Bring your left arm behind your back, with the elbow bent at 90 degrees.

3. Have a partner stand behind you facing your back and grasp the left elbow.

4. The partner gently pulls the elbow back and up toward the head, taking care not to pull suddenly or with great force.

5. Repeat these steps for the opposite arm.

Muscles Stretched

Most-stretched muscles: Left supraspinatus, left infraspinatus

Less-stretched muscles: Left anterior deltoid, left pectoralis major, left teres minor, left coracobrachialis

Stretch Notes

The supraspinatus and infraspinatus muscles can become tight when a person does either repeated forward pushing actions, as when using a walk-behind lawn mower, or downward pulling actions, such as raising something off the ground using a block-and-tackle pulley system. The supraspinatus especially is always working during overhead movements and so can be easily strained when it fatigues. This stretch can also help relieve the pain associated with shoulder impingement, shoulder bursitis, rotator cuff tendinitis, and frozen shoulder.

If you have ever had someone twist your arm behind your back, you know that this movement can be very painful. The pain is magnified if these muscles are very tight. Therefore, the person assisting with this stretch needs to proceed slowly when pulling the arm up and back.

ARMS, WRISTS, AND HANDS

The major joint of the arm, the elbow, is made up of three bones. The humerus (upper arm) is located proximal to the body while the radius and ulna (forearm) lie distally. The elbow is a hinge and thus has only the capacity to either flex or extend. As a result, the muscles that flex the elbow (biceps brachii, brachialis, brachioradialis, pronator teres) are located anteriorly (on the front; figure 3.1), whereas the extensor muscles (anconeus, triceps brachii) are located posteriorly (on the back; figure 3.2).

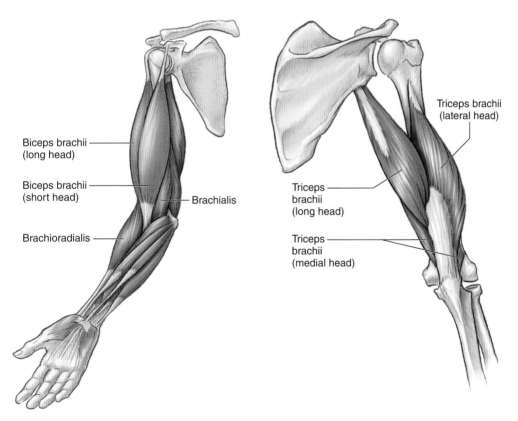

Biceps brachii (long head)

Biceps brachii (short head)

Brachialis

Brachioradialis

Triceps brachii (lateral head)

Triceps brachii (long head)

Triceps brachii (medial head)

Figure 3.1 Biceps brachii, brachialis, and brachioradialis muscles.

Figure 3.2 Triceps brachii muscle.

The ligaments that help hold the three bones of the elbow joint in place are the joint capsule ligament, the radial collateral ligament, and the ulnar collateral ligament. The radius gets its name from its ability to roll over the ulna, and this ability allows the palm to face either forward (supinated) or backward (pronated). The head of the radius is connected to the ulna via the annular ligament. There are two muscles that supinate (biceps brachii and supinator) and two muscles that pronate (pronator teres and pronator quadratus). The pronator muscles are located so they can pull the distal radius toward the center of the body, and the supinator muscles are situated to pull the distal radius away from the body.

The degree of available elbow flexion is limited primarily by the forearm contacting the anterior muscles of the upper arm, as well as the anterior proximal ends of the radius and ulna contacting the anterior distal end of the humerus. The tightness of the elbow extensors, however, along with the strength of the elbow flexors and the flexibility of the posterior portions of the capsular, radial collateral, and ulnar collateral ligaments also control the range of movement. These can be altered by stretching.

Although the major movements at the wrist are flexion and extension, the wrist is a gliding joint and not a true hinge joint. The gliding is possible because the wrist consists of the distal ends of the radius and ulna and the eight wrist, or carpal, bones. Thus, in addition to flexion and extension, the wrist can perform abduction (radial deviation) and adduction (ulnar deviation). The carpal bones are mostly held together by the various joint capsules, the palmar radiocarpal ligament, and the dorsal radiocarpal ligament. Interestingly, most of the muscles that control wrist, hand, and finger movements are located at or near the elbow. This results in the belly of the muscle lying near the elbow, with tendons crossing the wrist and attaching to the wrist bones (carpals), hand bones (metacarpals), and finger bones (phalanges). Having only tendons in the wrists and hands prevents the wrists and hands from getting too bulky from the increase in size that accompanies muscle strength.

Similar to the muscles that move the elbow, all the wrist flexors (flexor carpi radialis, flexor carpi ulnaris, and palmaris longus) and most of the finger flexors (flexor digitorum profundus, flexor digitorum superficialis, and flexor pollicis longus) are located in the anterior compartment of the forearm (figure 3.3a). In contrast, all the wrist extensors (extensor carpi radialis brevis, extensor carpi radialis longus, extensor carpi ulnaris, extensor digitorum communis) and finger extensors (extensor digitorum communis, extensor digiti minimi, extensor indicis) are located in the posterior compartment of the forearm (figure 3.3b). The muscles that run along the radius, which have radialis in their names, perform ulnar deviation, or wrist abduction. Those along the ulna, which have ulnaris in their names, perform radial deviation, or wrist adduction. Just before crossing the wrist, the tendons of these muscles are anchored firmly by thick tissue bands called the flexor retinaculum and extensor retinaculum. By passing under the retinaculum at the carpals (wrist bones), the tendons lie in a carpal tunnel. Since the tendons are crowded together, each tendon is surrounded by a slippery sheath to minimize friction.

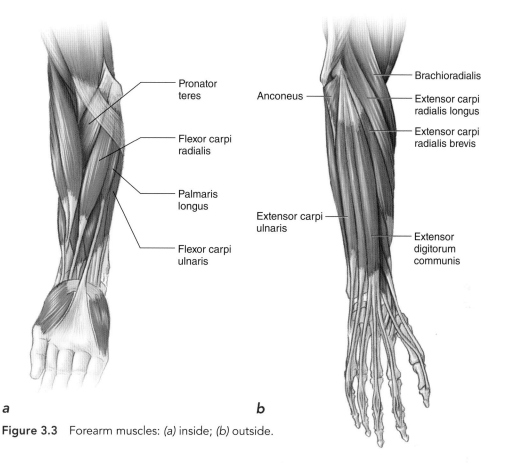

Figure 3.3   Forearm muscles: *(a)* inside; *(b)* outside.

The movement ranges for wrist flexion, wrist extension, radial deviation, and ulnar deviation are all limited by the strength of the agonist muscles, flexibility of the antagonist muscles, tightness of the dorsal and palmar ligaments, and wrist impingement (ulnar deviation only). Interestingly, all of these, except wrist impingement, can be changed by doing stretching exercises.

Stretching the muscles that move the elbows and wrists is helpful in alleviating and sometimes preventing overuse injuries. Because it is more resistive to opposing movements, a tight muscle is easy to damage. When the wrist extensor muscles are tight, pain arises on the lateral (outer) side of the elbow. In sports, this pain is sometimes referred to as tennis elbow. Tight wrist flexor muscles, on the other hand, can cause pain on the opposite, or medial, side of the elbow. This pain is frequently called golfer's elbow. Also, tightness in both the wrist extensors and flexors from either constant wrist hyperextension or flexion can lead to increased friction, inflammation, and overuse injuries such as carpal tunnel syndrome. People engaged in static or fine motor work such as keyboard use, computer mouse use, carpentry, or rock climbing are most likely to encounter this condition. To prevent and alleviate this condition, rehabilitation specialists encourage work breaks for stretching both the wrist flexors and extensors to help strengthen and loosen the muscles and tendons.

Many of the instructions and illustrations in this chapter are given for the left side of the body. Similar but opposite procedures would be used for the right side of the body. The stretches in this chapter are excellent overall stretches for all the arm muscles. However, some people may need to target a specific muscle or group and, hence, require stretches more suited to their needs. Specific muscle stretches require the involvement of one or more movements in the opposite direction of the desired muscle's movements. For example, if you want to stretch the flexor carpi radialis, perform a movement that involves wrist extension and radial deviation. When a muscle has a high level of stiffness, however, you should use fewer simultaneous opposite movements. For example, to stretch a very tight flexor carpi radialis, start by doing only radial deviation. As a muscle becomes loose, you can then incorporate more simultaneous opposite movements.

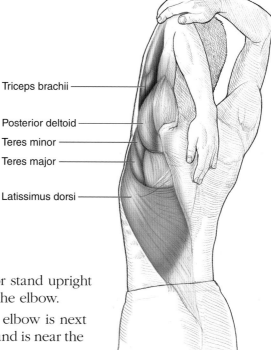

Triceps brachii

Posterior deltoid

Teres minor

Teres major

Latissimus dorsi

Execution

1. Sit in a chair with a back or stand upright with the left arm flexed at the elbow.
2. Raise the left arm until the elbow is next to the left ear and the left hand is near the right shoulder blade.
3. Grasp the upper arm just below the left elbow with the right hand, and pull or push the left elbow behind the head and toward the floor.
4. Repeat these steps for the opposite arm.

Muscles Stretched

Most-stretched muscle: Left triceps brachii

Less-stretched muscles: Left latissimus dorsi, left teres major, left teres minor, left posterior deltoid

Stretch Notes

Tightness in the elbow extensor muscles is the main cause of tennis elbow, or pain in the lateral elbow during arm movements. This tightness is usually caused by overworking or straining these muscles or by working against resistance with the arm fully extended. Therefore, any activity that uses these muscles can lead to tightness. Consequently, this stretch is beneficial not only for tennis players but also for swimmers. Alternatively, strain can result if the muscle is constantly overstretched by tight elbow flexors or if the arm is muscle bound (inability to completely straighten the arm).

Doing this stretch while seated in a chair with a back allows better control of balance. A greater stretching force can be applied to the muscles when the body is balanced. Also, do not perform this stretch for an extended period because this stretch greatly reduces blood flow to the shoulder.

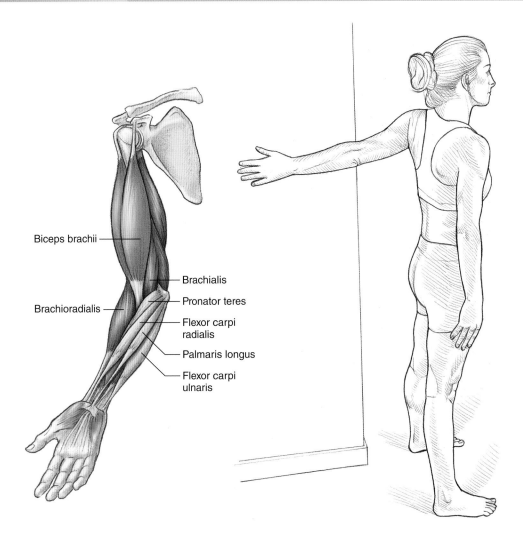

Biceps brachii

Brachialis

Pronator teres

Brachioradialis

Flexor carpi radialis

Palmaris longus

Flexor carpi ulnaris

Execution

1. Stand facing the inside of a doorframe, but at an arm's length.
2. Raise the left arm to shoulder level, keeping it straight.
3. Grasp the farthest edge of the doorframe, with the thumb pointing up.
4. Keeping the left elbow and wrist straight, rotate the trunk back toward the doorframe.
5. Repeat these steps for the opposite arm.

Muscles Stretched

Most-stretched muscles: Left brachialis, left brachioradialis, left biceps brachii

Less-stretched muscles: Left supinator, left pronator teres, left flexor carpi radialis, left flexor carpi ulnaris, left palmaris longus

Stretch Notes

These flexor muscles easily become tight from large amounts of bent elbow work such as carrying heavy boxes or curling either dumbbells or barbells. When these muscles are tight, the arm cannot be completely straightened, and the person has what is often called a muscle-bound look. This tightness causes pain on the medial elbow, often referred to as golfer's elbow. However, the pain is not limited to golfers and can affect other people such as carpenters, rock climbers, massage therapists, and weightlifters. Also, stretching these flexor muscles can bring relief to those who suffer from carpal tunnel syndrome.

This stretch is easier to do if you grasp a solidly fixed vertical pole. Grasp the pole firmly so your hand does not slide along the pole, but do not grasp too tightly as a tight grasp virtually eliminates the stretch effect on the less-stretched muscles. Also, it is more difficult to keep the elbow straight, and a straight elbow is necessary for this stretch to be effective. It is preferable to lift the arm to shoulder level to ensure that all muscles receive the same amount of stretch. Nevertheless, the stretch will be effective at whatever height the arm is raised.

ELBOW AND WRIST FLEXOR STRETCH

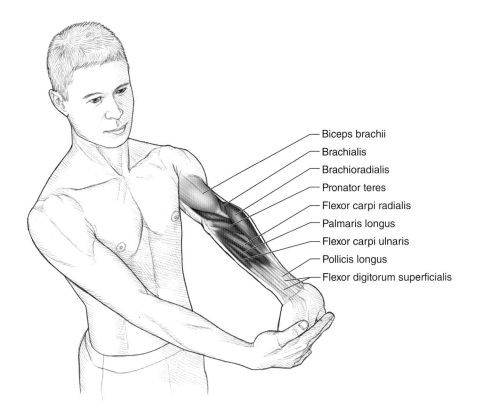

- Biceps brachii
- Brachialis
- Brachioradialis
- Pronator teres
- Flexor carpi radialis
- Palmaris longus
- Flexor carpi ulnaris
- Pollicis longus
- Flexor digitorum superficialis

Execution

1. Stand upright with feet shoulder-width apart, toes pointing straight forward.
2. Stick your left arm out in front of you at shoulder height, with the elbow straight and forearm supinated (turned up).
3. Hyperextend the left wrist so that the fingers point toward the floor.
4. Grab the left fingers with the right hand, and pull the fingers back toward the elbow.
5. Repeat these steps for the opposite arm.

Muscles Stretched

Most-stretched muscles: Left brachialis, left brachioradialis, left pronator teres, left flexor carpi radialis, left flexor carpi ulnaris, left palmaris longus

Less-stretched muscles: Left biceps brachii, left flexor digitorum superficialis, left flexor digitorum profundus, left pollicis longus

Stretch Notes

These flexor muscles easily become tight from static work such as operating a keyboard. Also any occupation that requires high amounts of arm work can cause these muscles to become tight. This tightness causes pain on the medial elbow, often referred to as golfer's elbow. However, the pain is not limited to golfers and can affect other people such as carpenters, rock climbers, and massage therapists. Also, stretching these flexor muscles can help bring relief to those with carpal tunnel syndrome.

Exercise caution when doing this stretch. If you feel any pain in the elbow, wrist, or finger joints, reduce the tension or joint damage could occur.

ANCONEUS STRETCH

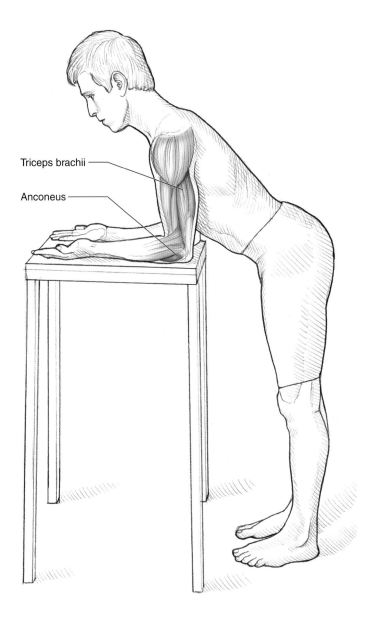

Triceps brachii

Anconeus

Execution

1. Stand or sit upright while facing a table that is about waist high.
2. Flex the elbows and rest the forearms on the table with the palms up.
3. Lean forward, bringing the chest toward the table.

Muscles Stretched

Most-stretched muscle: Anconeus

Less-stretched muscle: Triceps brachii

Stretch Notes

Tightness in the elbow extensor muscles is the main cause of tennis elbow, or pain in the lateral elbow during arm movements. This tightness usually is caused by overworking or straining these muscles. Therefore any activity that uses these muscles can lead to tightness. Although the triceps brachii is the major muscle used in extending the elbow, the anconeus becomes a major player when the arm is bent and pronated. Hence, tennis players who mainly use a close-to-the-body forehand stroke or someone who has the muscle-bound look (unable to straighten the arms) will benefit greatly from this stretch.

For the greatest stretch, keep the forearms and elbows flat on the table.

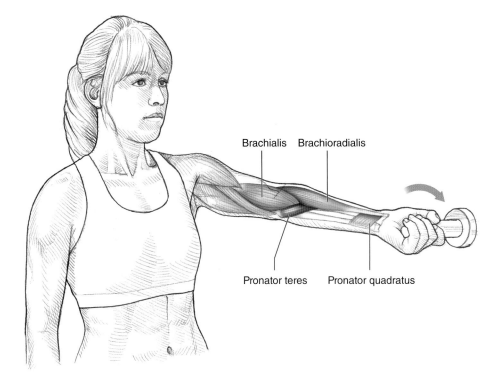

Brachialis   Brachioradialis

Pronator teres   Pronator quadratus

Execution

1. Stand upright with feet shoulder-width apart, toes pointing straight forward.
2. With the left hand, grasp a light dumbbell with a weight plate attached to one end only, with the weighted end sticking out past the thumb.
3. Stick your left arm out in front of you at shoulder height, with the elbow straight and forearm supinated (the top of the weight left of the thumb).
4. Hypersupinate the forearm (rotate the wrist toward the thumb) so that the weighted end of the dumbbell points toward the floor.
5. Repeat these steps for the opposite arm.

Muscles Stretched

Most-stretched muscle: Left pronator teres

Less-stretched muscles: Left brachialis, left brachioradialis, left pronator quadratus

Stretch Notes

A pronation contracture, or extremely tight pronator muscles, is primarily caused by hypertonicity (a shortened, stiff muscle) in the pronator teres. This hypertonicity can cause medial nerve compression, or pronator teres syndrome. The symptoms are felt as pain and weakness in the anterior forearm and hand. Pronator teres syndrome results from overuse of the pronator teres through repetitive occupational activities such as hammering, cleaning fish, or performing any activity that requires continual manipulation of tools. Women are affected more than men, although the reason for this is not clear. Regularly stretching the pronator teres can help reduce the possibility of developing contractures.

Be careful not to use a weight that is too heavy. Start with a very light weight plate on one end of the dumbbell, and gradually increase the weight as you become more used to the stretch. In fact, you do not need to use a dumbbell at all. Any object that has a light weight on one end of a handle, such as a hammer, will work just as well. Also, this stretch can be done either sitting or standing, with the whole arm lying on a flat surface and the wrist and hand extended past the edge of the surface. If you do use a support, try to keep the shoulder angle near 90 degrees.

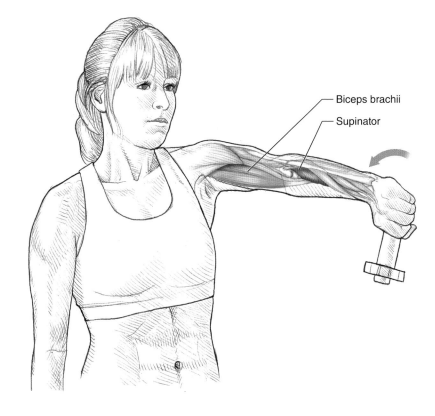

Biceps brachii

Supinator

Execution

1. Stand upright with your feet shoulder-width apart, toes pointing straight forward.

2. In your left hand, grasp a light dumbbell with a weight plate attached to one end only, with the weighted end sticking out past the thumb.

3. Stick your left arm out in front of you at shoulder height, with the elbow straight and forearm supinated.

4. Pronate the forearm (rotate the wrist toward the little finger) so that the weighted end of the dumbbell points toward the floor.

5. Repeat these steps for the opposite arm.

Muscles Stretched

Most-stretched muscle: Left supinator

Lesser-stretched muscle: Left biceps brachii

Stretch Notes

A short and tight (hypertonic) supinator is a major contributor to lateral elbow pain, often called tennis elbow. A severe hypertonic supinator can contribute to either supinator syndrome or radial tunnel syndrome. These syndromes are the result of radial nerve compression and manifest themselves as forearm pain and numbness along with weakness in the lower arm and hand muscles. Movements such as a quick tennis backhand or prolonged forearm supination with a flexed elbow, such as cutting hair, walking a dog on a leash, or carrying heavy boxes from underneath, are the types of movements that can overwork the supinator and lead to a hypertonic muscle.

Be careful not to use a weight that is too heavy. Start with a very light weight plate on one end of the dumbbell, and gradually increase the weight as you become more used to the stretch. In fact, you do not need to use a dumbbell at all. Any object that has a weight on one end of a handle, such as a hammer, will work just as well. Also, this stretch can be done either sitting or standing, with the whole arm lying on a flat surface and the wrist and hand extended past the edge of the surface. If you do use a support, try to keep the shoulder angle near 90 degrees.

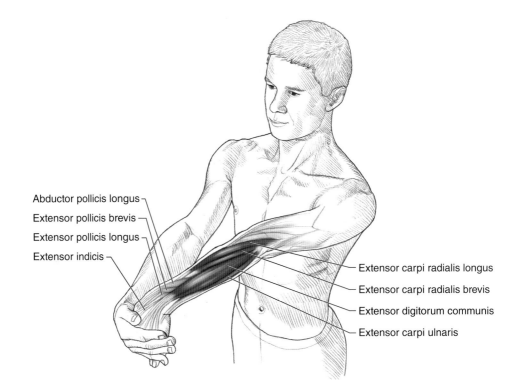

Abductor pollicis longus
Extensor pollicis brevis
Extensor pollicis longus
Extensor indicis

Extensor carpi radialis longus
Extensor carpi radialis brevis
Extensor digitorum communis
Extensor carpi ulnaris

Execution

1. Stand upright with feet shoulder-width apart, toes pointing straight forward.
2. Stick your left arm out in front of you at shoulder height, with the elbow straight and forearm pronated.
3. Bend the left wrist so that the fingers point toward the floor.
4. Place the palm of the right hand against the knuckles of the left hand.
5. While keeping the left elbow straight, pull the knuckles toward the body.
6. Repeat these steps for the opposite arm.

Muscles Stretched

Most-stretched muscles: Left extensor carpi radialis brevis, left extensor carpi radialis longus, left extensor carpi ulnaris, left extensor digitorum communis

Less-stretched muscles: Left extensor indicis, left extensor pollicis brevis, left extensor pollicis longus, left abductor pollicis longus

Stretch Notes

Tightness in the extensor muscles is a cause of tennis elbow, or pain in the lateral elbow during arm movements. This tightness usually is caused by overworking or straining these muscles. Any activity that uses these muscles, such as keyboard work, racket sports, rowing, weightlifting, wheelchair sports, and rock climbing, can lead to overwork, hypertonicity, and tightness. Also, overworking the extensor pollicis longus and brevis or the abductor pollicis longus can lead to conditions known as drummer boy palsy (mainly the extensor pollicis longus) and intersection syndrome (mainly the extensor pollicis brevis and abductor pollicis longus). By doing this stretch, you help reduce the problems that can arise from overworked wrist extensors.

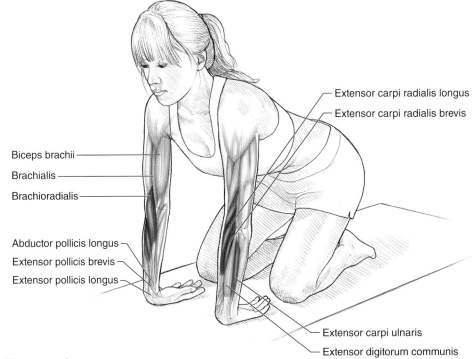

Extensor carpi radialis longus

Extensor carpi radialis brevis

Biceps brachii

Brachialis

Brachioradialis

Abductor pollicis longus

Extensor pollicis brevis

Extensor pollicis longus

Extensor carpi ulnaris

Extensor digitorum communis

Execution

1. Kneel on the floor.
2. Flex both wrists and place the backs of your hands on the floor, hands shoulder-width apart.
3. Point the fingers toward the knees.
4. While keeping the elbows straight, lean back, bringing the buttocks to the heels, keeping the backs of the hands on the floor.

Muscles Stretched

Most-stretched muscles: Brachioradialis, extensor carpi radialis brevis, extensor carpi radialis longus, extensor carpi ulnaris

Less-stretched muscles: Supinator, brachialis, biceps brachii, extensor digitorum communis, extensor pollicis brevis, extensor pollicis longus, abductor pollicis longus

Stretch Notes

Tightness in the extensor muscles can cause tennis elbow, or pain in the lateral elbow during arm movements. This tightness usually is caused by overworking or straining these muscles. Therefore any activity that uses these muscles, such as keyboard work, racket sports, rowing, weightlifting, wheelchair sports, and rock climbing, can lead to overwork, hypertonicity, and tightness. Overworking the extensor pollicis longus and brevis or the abductor pollicis longus can lead to conditions known as drummer boy palsy (mainly the extensor pollicis longus) and de Quervain syndrome (mainly the extensor pollicis brevis and abductor pollicis longus). The beginner stretch is best for those who have a small range of wrist motion or severe pain when using the wrist. Once you gain more range of motion, however, you should do this intermediate stretch to reduce the problems that can arise from overworked wrist extensors. This stretch will also strengthen the afflicted muscles and start you on the road to prevention of further problems.

The closer the hands are to the knees, the easier it is to keep the backs of the hands touching the floor. The farther the hands are in front of the knees, however, the greater the applied stretch.

⟨ **VARIATION** ⟩

Wrist Radial Deviator and Extensor Stretch

By changing the direction that the fingers are pointing, you can change the stretching emphasis on the forearm muscles. For instance, both the wrist extensor muscles and the radial deviator muscles can be stretched simultaneously. First, assume the starting position by kneeling on the floor with the wrists flexed and the backs of your hands on the floor. Second, instead of pointing the fingers toward the knees, rotate the hands so that the fingers point medially (fingertips point toward each other). Finally, stretch the desired muscles by leaning back (buttocks to the heels) while keeping the backs of the hands on the floor.

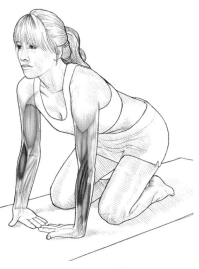

⟨ **VARIATION** ⟩

Wrist Ulnar Deviator and Extensor Stretch

If you change the direction the fingers point, you alter the stretching emphasis on the forearm muscles. To stretch both the wrist extensor muscles and the ulnar deviator muscles simultaneously, first assume the starting position by kneeling on the floor with the wrists flexed and the backs of your hands on the floor. Second, instead of pointing the fingers toward the knees, rotate the hands so the fingers point laterally (fingertips point away from the body on a line perpendicular to the midline of the body). Finally, stretch the desired muscles by leaning back (buttocks to the heels) while keeping the backs of the hands on the floor.

ARMS · WRISTS · HANDS

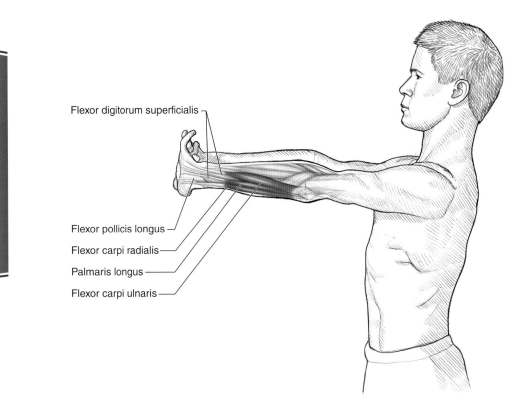

Flexor digitorum superficialis

Flexor pollicis longus

Flexor carpi radialis

Palmaris longus

Flexor carpi ulnaris

Execution

1. Stand upright with feet shoulder-width apart, toes pointing straight forward.

2. Interlock your fingers, with the palms pointing out away from the body.

3. With the arms at shoulder level, straighten your elbows and push your palms out away from the body as far as you can.

Muscles Stretched

Most-stretched muscles: Flexor carpi radialis, flexor carpi ulnaris, pronator teres, palmaris longus

Less-stretched muscles: Flexor pollicis longus, flexor digitorum profundus, flexor digitorum superficialis

Stretch Notes

The flexor muscles easily become tight from repeated use of the arm or wrist in an awkward position or by bending the wrist while typing, using the phone, or operating a machine. Additional problems arise from working with the arm held away from the body or playing sports. This tightness causes pain on the medial elbow, often referred to as golfer's elbow. The longer a person does any of these activities, the greater the risk of tightness and the greater the need to stretch these muscles.

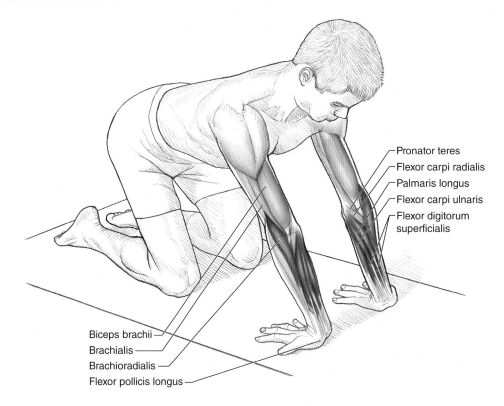

Pronator teres
Flexor carpi radialis
Palmaris longus
Flexor carpi ulnaris
Flexor digitorum superficialis

Biceps brachii
Brachialis
Brachioradialis
Flexor pollicis longus

Execution

1. Kneel on the floor.
2. Flex both wrists and place the palms of your hands on the floor, hands shoulder-width apart.
3. Point the fingers toward the knees.
4. While keeping the elbows straight, lean back (buttocks to the heels), keeping the palms flat on the floor.

Muscles Stretched

Most-stretched muscles: Brachioradialis, flexor carpi radialis, flexor carpi ulnaris, flexor digitorum profundus, flexor digitorum superficialis, palmaris longus

Less-stretched muscles: Flexor digiti minimi brevis, flexor pollicis longus, pronator teres, brachialis, biceps brachii

Stretch Notes

The flexor muscles easily become tight from repeated use of the arm or wrist in an awkward position or by bending the wrist while typing, using the phone, or operating a machine. Additional problems arise from working with the arm held away from the body or playing sports. This tightness causes pain on the medial elbow, often referred to as golfer's elbow. The longer a person does any of these activities, the greater the risk of tightness and the greater the need for stretching these muscles. Unfortunately, the beginner-level exercise provides only limited stretch. As you increase your flexibility, you need to move to a more intense stretch, such as this intermediate one.

The closer the hands are to the knees, the easier it is to keep the palms of the hands touching the floor. The farther the hands are away from the midline, the greater the stretch.

⟨ **VARIATION** ⟩

Wrist Radial Deviator and Flexor Stretch

If you change the direction that the fingers are pointing, you alter the stretching emphasis being placed on the forearm muscles. To stretch both the wrist flexor muscles and the radial deviator muscles simultaneously, assume the starting position by kneeling on the floor, with the wrists flexed and the palms of your hands on the floor. Second, instead of pointing the fingers toward the knees, rotate the hands so the fingers point laterally (fingertips point away from the body on a line perpendicular to the midline of the body). Finally, stretch the desired muscles by leaning back (buttocks to the heels) while keeping the palms of the hands on the floor.

⟨ **VARIATION** ⟩

Wrist Ulnar Deviator and Flexor Stretch

By changing the direction the fingers are pointing, you change the stretching emphasis on the forearm muscles. For instance, both the wrist flexor muscles and the ulnar deviator muscles can be stretched simultaneously. First, assume the starting position by kneeling on the floor, with the wrists flexed and the palms of your hands on the floor. Second, instead of pointing the fingers toward the knees, rotate the hands so the fingers point medially (fingertips point toward each other). Finally, stretch the desired muscles by leaning back (buttocks to the heels) while keeping the palms of the hands on the floor.

WRIST RADIAL DEVIATOR STRETCH WITH DUMBBELL

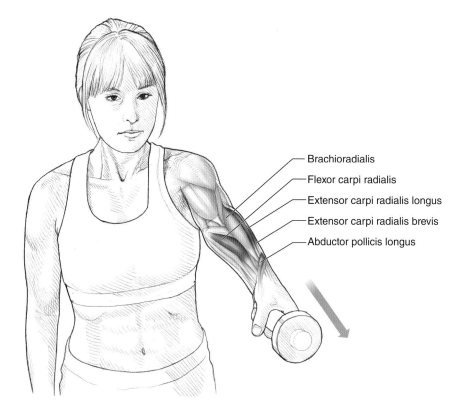

Brachioradialis
Flexor carpi radialis
Extensor carpi radialis longus
Extensor carpi radialis brevis
Abductor pollicis longus

Execution

1. Stand upright with feet shoulder-width apart, toes pointing straight forward.

2. In your left hand, grasp a dumbbell with a weight plate attached to one end only, with the weighted end sticking out past the thumb.

3. Stick your left arm out in front of you at shoulder height, with the elbow straight and forearm rotated so that the thumb side of the hand points up.

4. Bend the wrist down so that the weighted end of the dumbbell points more forward, away from the body rather than up.

5. Repeat these steps for the opposite arm.

Muscles Stretched

Most-stretched muscles: Left abductor pollicis longus, left flexor carpi radialis, left extensor carpi radialis longus, left extensor carpi radialis brevis

Less-stretched muscles: Left brachioradialis

Stretch Notes

Many activities that require using the wrist in repetitive actions for many hours each day, such as extended work on a computer or tennis, golf, baseball, bowling, and mountain biking, force the wrist joint to the extremes of its range of motion and make the area vulnerable to tightness or hypertonicity. If done without adequate rest and recovery, the limited, repetitive motions involved in playing the violin or piano can also cause tightness. Also, the wrist may be damaged in simple, mundane activities such as scrubbing a pot, pushing up out of a chair, or lifting a small object in an awkward position. Much of the tightness, pain, and injury associated with these activities can be relieved through stretching the wrist radial deviators.

Be careful not to use a weight that is too heavy. Start with a very light weight plate on one end of the dumbbell, and gradually increase the weight as you become more used to the stretch. In fact, to do this stretch you do not need to use a dumbbell at all. Any object with a weight attached to one end of a handle, such as a hammer, will work just as well. Also, this stretch can be done either sitting or standing, with the whole arm lying on a flat surface and the wrist and hand extended past the edge of the surface. If you do use a support, try to keep the shoulder angle near 90 degrees.

WRIST ULNAR DEVIATOR STRETCH WITH DUMBBELL

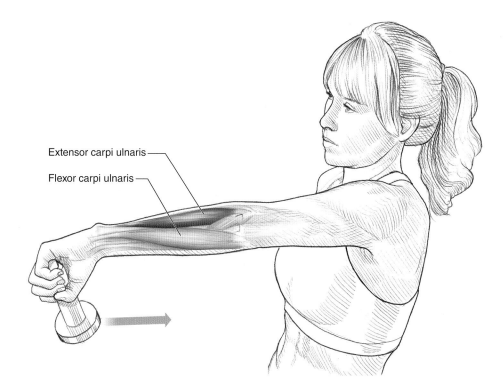

Extensor carpi ulnaris

Flexor carpi ulnaris

Execution

1. Stand upright with feet shoulder-width apart, toes pointing straight forward.
2. In the left hand, grasp a dumbbell with a weight plate attached to one end only, with the weighted end sticking out past the thumb.
3. Stick your left arm out in front of you at shoulder height, with the elbow straight and forearm rotated so that the thumb side of the hand points down.
4. Bend the wrist down so that the weighted end of the dumbbell points more toward the body rather than down.
5. Repeat these steps for the opposite arm.

Muscles Stretched

Most-stretched muscle: Left extensor carpi ulnaris

Less-stretched muscle: Left flexor carpi ulnaris

Stretch Notes

Many activities that require using the wrist in repetitive actions for many hours each day, such as extended work on a computer or tennis, golf, baseball, bowling, and mountain biking, force the wrist joint to the extremes of its range of motion and make the area vulnerable to tightness or hypertonicity. If done without adequate rest and recovery, the limited, repetitive motions involved in playing the violin or piano can also cause tightness. Also, the wrist may be damaged in simple, mundane activities such as scrubbing a pot, pushing up out of a chair, or lifting a small object in an awkward position. Much of the tightness, pain, and injury associated with these activities can be relieved through stretching the wrist ulnar deviators.

Do not use a weight that is too heavy. Start with a very light weight plate on one end of the dumbbell, and gradually increase the weight as you become more used to the stretch. In fact, to do this stretch you do not need to use a dumbbell at all. Any object that has a weight attached to one end of a handle, such as a hammer, will work just as well. Also, this stretch can be done either sitting or standing, with the whole arm lying on a flat surface and the wrist and hand extended past the edge of the surface. If you do use a support, try to keep the shoulder angle near 90 degrees.

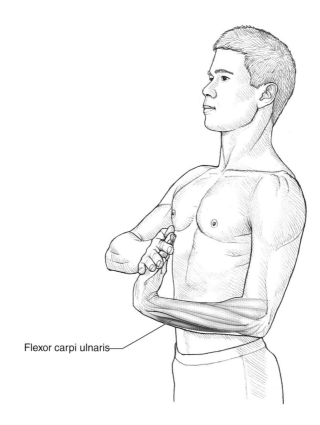

Flexor carpi ulnaris

Execution

1. Sit or stand upright.
2. Flex the elbow at a 90-degree angle, and extend the wrist as far as possible.
3. Point the fingers upward.
4. With the right hand, push the fingers on the left hand toward the elbow.
5. Repeat these steps for the opposite arm.

Muscles Stretched

Most-stretched muscles: Left flexor carpi radialis, left flexor carpi ulnaris, left flexor digiti minimi brevis, left flexor digitorum profundus, left flexor digitorum superficialis, left palmaris longus

Less-stretched muscle: Left flexor pollicis longus

Stretch Notes

Tightness and hypertonicity of the finger flexors usually arise from making a fist or curling the wrists into flexion. Sleeping with the hands in this position causes the flexor muscle group to become even tighter and shorter; causing impingement and damage to the median nerve within the carpal tunnel. The finger flexors also become tight from repetitive work in which the hand is grasping something for a long period of time, such as when hammering or rock climbing. A person can also develop what is called trigger finger by overworking the index finger. Also, some of the problems of the forearm such as golfer's elbow, or medial epicondylitis, are the result of tight finger flexor muscles. Finally, improper hand position in piano playing—wrist not relaxed, using a pushing action rather than a freely rebounding gravity stroke for the key stroke—can lead to finger flexor stiffness.

The elbow angle does not need to be precisely 90 degrees. Choose a comfortable angle. Some people find that fully flexing the elbow makes it easier to push on the hand. With the elbow fully flexed, the push is more down than across.

WALL-ASSISTED FINGER FLEXOR STRETCH

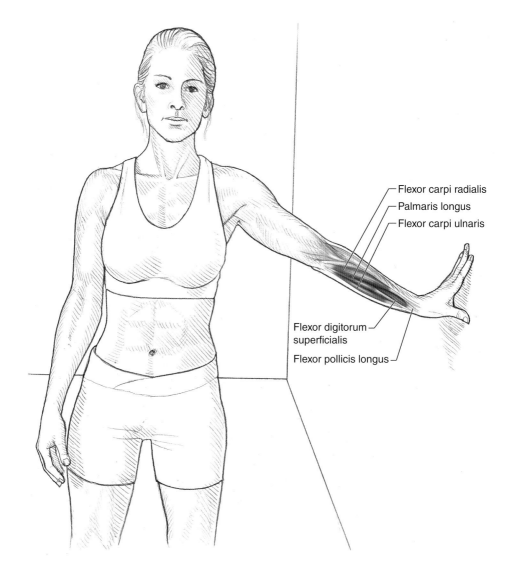

Flexor carpi radialis
Palmaris longus
Flexor carpi ulnaris

Flexor digitorum superficialis
Flexor pollicis longus

Execution

1. Stand upright about one foot (30 cm) from a wall.
2. Turn the body so that the left shoulder is perpendicular to the wall.
3. Reach out and place the left fingertips on the wall midway between the left hip and left shoulder.
4. While keeping only the left fingertips in contact with the wall, lean toward the wall.
5. Repeat these steps for the opposite arm.

Muscles Stretched

Most-stretched muscles: Left flexor carpi radialis, left flexor carpi ulnaris, left flexor digiti minimi brevis, left flexor digitorum profundus, left flexor digitorum superficialis, left palmaris longus

Less-stretched muscles: Left flexor pollicis longus

Stretch Notes

Tightness and hypertonicity of the finger flexors usually arise from making a fist or curling the wrists into flexion. Sleeping with the hands in this position causes the flexor muscle group to become even tighter and shorter, causing impingement and damage to the median nerve within the carpal tunnel. The finger flexors also become tight from repetitive work in which the hand is grasping something for a long period of time, such as when hammering or rock climbing. A person can also develop what is called trigger finger by overworking the index finger. Also, some of the problems of the forearm such as golfer's elbow, or medial epicondylitis, are the result of tight finger flexor muscles. Finally, improper hand position in piano playing—wrist not relaxed, with a pushing action rather than a freely rebounding gravity stroke as the key stroke—can lead to finger flexor stiffness.

The initial starting height of your fingers relative to the hips is not critical. You should start in a position that makes it easy to maintain balance while still putting stretch tension on the muscles. As you adapt to the stretch, you may find it necessary to change the finger height to achieve the desired stretching tension.

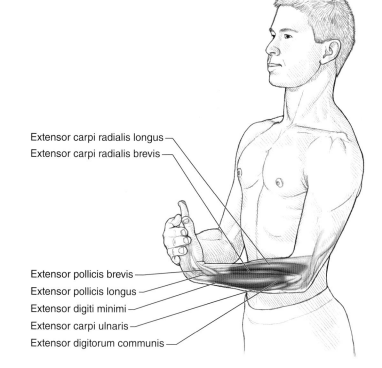

FINGER EXTENSOR STRETCH

Extensor carpi radialis longus
Extensor carpi radialis brevis

Extensor pollicis brevis
Extensor pollicis longus
Extensor digiti minimi
Extensor carpi ulnaris
Extensor digitorum communis

Execution

1. Sit or stand upright.
2. Turn the left arm so that the left palm faces up. Flex the left elbow to a 90-degree angle.
3. Flex the left wrist to a 90-degree angle. Flex the fingers so they point toward the elbow.
4. Place the right hand on top of the fingers, and press the fingers down toward the forearm.
5. Repeat these steps for the opposite arm.

Muscles Stretched

Most-stretched muscles: Left extensor carpi radialis brevis, left extensor carpi radialis longus, left extensor carpi ulnaris, left extensor digitorum communis, left extensor digiti minimi, left extensor indicis

Less-stretched muscles: Left extensor pollicis brevis, left extensor pollicis longus

Stretch Notes

Tightness in the extensor muscles is also a cause of tennis elbow, pain in the lateral elbow during arm movements. This tightness usually is caused by overworking or straining these muscles. Therefore any activity that uses these muscles, such as keyboard work, racket sports, rowing, weightlifting, wheelchair sports, and rock climbing, can lead to overwork, hypertonicity, and tightness. Also, overworking the extensor pollicis longus and brevis or the abductor pollicis longus can lead to conditions known as drummer boy palsy (mainly the extensor pollicis longus) and de Quervain syndrome (mainly the extensor pollicis brevis and abductor pollicis longus). A tight extensor carpi radialis longus or extensor carpi radialis brevis can also lead to inflammation of their respective tendons, which can lead to radial wrist pain or intersection syndrome. By doing this stretch, you help reduce the problems that can arise from overworked finger extensors. Finally, ability to do active finger extension is used as a reliable early predictor of recovery of arm function in stroke patients. Thus, stretching the finger extensor muscles after a stroke helps in the rehabilitation process.

Increase the magnitude of the stretch by flexing the fingers (i.e., make a fist). Also, the elbow angle does not need to be precisely 90 degrees. Choose a comfortable angle. Some people find that fully flexing the elbow makes it easier to push on the hand. With the elbow fully flexed, the push is more down than across.

chapter 4

LOWER TRUNK

The 12 thoracic vertebrae, 5 lumbar vertebrae, sacrum, ribs, and pelvic bones along with associated muscles and ligaments make up the flexible framework of the trunk. The vertebrae and the other bones, muscles, and ligaments work together to support and move the trunk. As in the neck, the vertebral bodies (the oval-shaped bones) of the trunk are connected by posterior and anterior ligaments along with other ligaments that connect each spinous and transverse (lateral bony protuberance) process to its corresponding part on the adjacent vertebrae. In addition, each vertebra is separated by an intervertebral disc. Compression of the vertebrae upon the discs allows the trunk to move forward, backward, and sideways, with the amount of movement limited in part by the vertebral facets.

The trunk movements are flexion (moving the chest and thighs toward each other), extension (moving the chest and thighs away from each other), hyperextension (moving the trunk back away from an erect position), lateral flexion and extension (shoulders tipped back and forth sideways), and rotation.

Since many of the muscles in the trunk come in right and left pairings, all these muscles are involved in lateral flexion, lateral extension, and rotation. For example, the right external oblique and internal oblique abdominal muscles help perform right lateral flexion, and the left external oblique and internal oblique abdominal muscles help perform right lateral extension. Several of the muscles involved in movements of the lower trunk run between the pelvic bones and either the spinal column or rib cage.

The external oblique, internal oblique, and rectus abdominis of the abdomen (figure 4.1) and quadratus lumborum (figure 4.2a) flex the trunk by pulling the rib cage toward the pelvis. The iliacus (figure 4.2b), a trunk flexor,

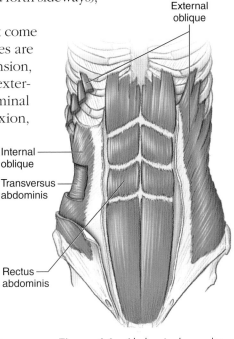

External oblique

Internal oblique

Transversus abdominis

Rectus abdominis

Figure 4.1 Abdominal muscles.

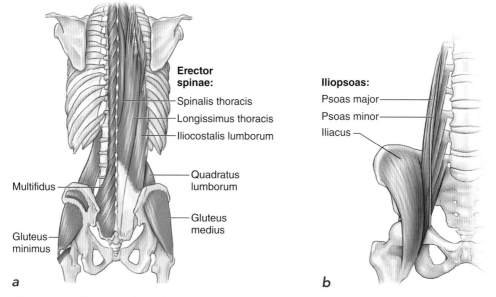

Figure 4.2   Core muscles: *(a)* posterior; *(b)* anterior.

pulls the femur (thigh bone) toward the pelvis. The psoas major, another trunk flexor, pulls the spinal column toward the femur. The prime trunk extensors (iliocostalis lumborum, longissimus thoracis, and spinalis thoracis) are collectively called the erector spinae. The iliocostalis lumborum runs between the posterior pelvis and posterior spinal column, while the longissimus thoracis and spinalis thoracis run along the posterior spinal column and help the individual vertebrae in the spinal column work together as a single unit. The interspinales, intertransversarii, multifidus, and rotatores run between individual vertebrae and cause large movements by making small changes between individual pairs or groups of vertebrae.

The ability to move the trunk is limited by the strength of the contracting muscles, the stiffness of the opposing ligaments, the stiffness of the noncontracting muscles, the alignment of the vertebral bodies with the adjacent vertebrae, the compressibility of the intervertebral discs, and the contact between body parts. For example, trunk flexion is limited by the stiffness of the posterior trunk muscles, the stiffness of the posterior trunk ligaments, the strength of the anterior trunk muscles, the alignment of the vertebral bodies with the adjacent vertebrae, the compressibility of the anterior portions of the intervertebral discs, the contact of the chin or rib cage with the legs, and the abdominal fat mass. Similarly, trunk extension is controlled by the stiffness of the anterior trunk muscles, the stiffness of the anterior trunk ligaments, the strength of the extensor muscles, the alignment of the vertebral bodies with the adjacent vertebrae, and the compressibility of the posterior portions of the intervertebral discs. In addition to the factors listed for flexion and extension, trunk lateral movement is controlled by the impingement of each vertebra's transverse process on the adjacent transverse processes. Trunk rotation is limited by the stiffness of spinal ligaments, the strength of the muscles on the side of rotation, the stiffness of the muscles opposite the side of rotation, and body tissues and their dimensions. For instance, rotating to the left is limited by weak left-side muscles and tight right-side muscles.

Many people who have stiff back muscles have discovered that stretching helps relieve some of the pain. The back muscles, or trunk extensors, are not the only lower-trunk muscles to influence back pain. Often people find relief from back pain by leaning back (trunk hyperextension) because this action stretches the abdominal muscles, the trunk flexors. This shows that flexible trunk flexors are also important. Moreover, numerous sporting activities such as golf, tennis, and throwing sports require twisting of the trunk. Twisting the trunk involves the trunk extensors, flexors, and lateral flexors. Improved range of motion of all lower-trunk muscles can increase the range of motion in trunk rotation and improve the performance in activities that involve these actions.

Hyperextension (arching) and hyperflexion (bending) of the lower back are potentially dangerous, especially if you have weak abdominal, thigh, and buttocks muscles. Backward rolling movements are potentially dangerous to the cervical spine (neck). Potential injuries include excessively squeezing the spinal discs, jamming together the spinal joints, and pinching the spinal nerves emerging from the lumbar vertebrae. If you choose to perform these stretches, build up to them more gradually than most other stretches. Also, to keep pressure off the neck during back rolls, keep the shoulder blades in contact with the floor.

Overstretching (very hard stretching) causes more harm than good. Sometimes the muscles become stiff from overstretching. Overstretching can reduce muscle tone, and the body compensates by making the loose muscle excessively tight. For each progression, start with the position that is the least stiff, and progress to the next position only when, after several days of stretching, you notice a consistent lack of stiffness during the exercise. This means you should stretch both the agonist and antagonist muscles. Also, remember that although there may be greater stiffness in one direction (right versus left), you should stretch both sides so you maintain proper muscle balance.

Many of the stretches in this chapter are described for the left side of the body. Similar but opposite procedures would be used for the right side of the body. The stretches in this chapter are excellent overall stretches. However, not all these stretches may be completely suited to each person's needs. To stretch specific muscles, the stretch must involve one or more movements in the opposite direction of the desired muscle's movements. For example, if you want to stretch the left external oblique, perform a movement that involves trunk extension and right trunk lateral flexion. When a muscle has a high level of stiffness, you should use fewer simultaneous opposite movements. For example, to stretch a very tight external oblique, start by doing only trunk extension. As a muscle becomes loose, you can incorporate more simultaneous opposite movements.

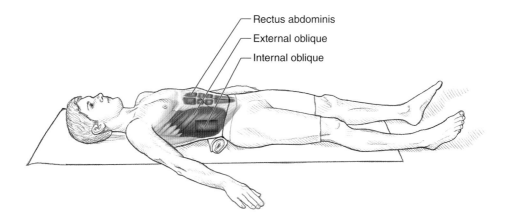

Rectus abdominis
External oblique
Internal oblique

Execution

1. Lie on the floor on your back.
2. Place a rolled-up towel (1 to 2 inches, or 2.5 to 5 cm, in diameter) between the small of your back and the floor.

Muscles Stretched

Most-stretched muscles: Rectus abdominis, external oblique, internal oblique

Less-stretched muscles: Quadratus lumborum, psoas major, iliacus

Stretch Notes

Although many people think that having tight abdominal muscles improves their overall appearance, tight abdominal muscles can have very negative effects on the body. First, tight abdominals are a major cause of lower-back pain. When tight, these muscles pull up on the pubic bones and tilt the top of the pelvis backward. Over time the upper-back muscles weaken and overlengthen, causing a flattening of the lumbar curve, which increases the pressure on the lumbar joints and discs. The constant stretch and compression of the discs result in chronic pain. In addition, when these muscles are tight, the volume of the abdominal and pelvic cavities is reduced. This compresses the organs in these cavities and forces them up toward the thoracic cavity, which in turn reduces its volume. As a result, breathing, digestion, elimination, and sexual function are hindered from functioning properly. Finally, exercising with tight abdominals can lead to strains, tears, and even hernias.

This stretch is especially recommended for people who have a swayed back or weak abdominal muscles, since arching the lower back is potentially dangerous for these people. Because the small of the back is supported in this exercise, undesired pressures on the spinal column are reduced. Nevertheless, the width of the back support is important. The larger the diameter of the towel, the greater the undesired pressure. Make sure the upper back, shoulder blades, and buttocks are resting comfortably on the floor. Also, squeezing the buttocks will reduce stress on the lower back.

PRONE LOWER-TRUNK FLEXOR STRETCH

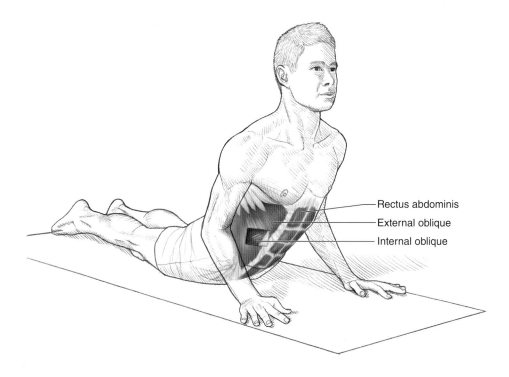

Rectus abdominis

External oblique

Internal oblique

Execution

1. Lie prone (facedown) on the floor.
2. Place both hands palms down. Fingers point forward by each hip.
3. Slowly arch the back, contracting the buttocks.
4. Continue arching the back as you lift your head and chest off the floor without hunching the shoulders.

Muscles Stretched

Most-stretched muscles: Rectus abdominis, external oblique, internal oblique

Less-stretched muscles: Quadratus lumborum, psoas major, iliacus, rotatores, intertransversarii

Stretch Notes

Those who spend a lot of time driving or sitting at a desk tend to slump forward, rounding the upper back, which also tightens the abdominal muscles. Having tight abdominal muscles is equivalent to wearing a corset. This compression of the abdominal and pelvic cavities can cause deterioration of the back muscles, restrict breathing, and interfere with the working of the viscera. When these muscles are tight, the diaphragm cannot go down and the rib cage cannot expand. Poor respiration can result in chronic fatigue, depression, asthma, and other consequences caused by inadequate oxygenation of the blood. Also, the organs in the abdominal cavity do not work well in a confined space. Kidney and bladder functions may be reduced. The uterus can be forced downward, increasing pressure and reducing blood flow. Increased pressure and decreased blood flow to the prostate may occur.

Remember that arching the lower back is potentially dangerous, especially if you have weak abdominal muscles. Injuries from arching the lower back include excessive squeezing of the spinal discs, jammed spinal joints, and pinched spinal nerves emerging from the lumbar vertebrae. Therefore, this stretch is recommended only for those who are very stiff. When doing this stretch, do minimal arching and make sure you squeeze the buttocks during the arching. Squeezing the buttocks reduces stress on the lower back.

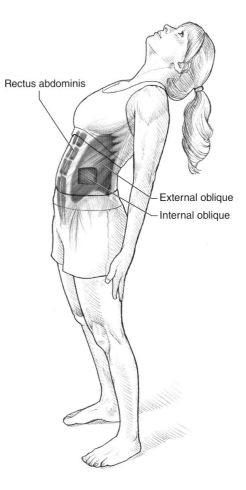

Rectus abdominis

External oblique

Internal oblique

Execution

1. Stand upright with the legs 2 to 3 feet (60 to 90 cm) apart, with hands on the hips.
2. Slowly arch the back, contracting the buttocks and pushing the hips forward.
3. As you continue to arch the back, drop the head back and slide the hands past the buttocks and down the legs.

Muscles Stretched

Most-stretched muscles: Rectus abdominis, external oblique, internal oblique

Less-stretched muscles: Quadratus lumborum, psoas major, iliacus

Stretch Notes

This exercise is potentially dangerous, especially for those who have a swayed back or weak abdominal muscles. This exercise can worsen a swayed back and cause excessive squeezing of the spinal discs, jammed spinal joints, and pinched spinal nerves emerging from the lumbar vertebrae. This stretch is recommended only for those who are very stiff and do not have a swayed back. Also, use this exercise only when the other lower-back flexor stretches do not provide any improvement. When doing this stretch, do minimal arching and make sure you squeeze the buttocks during the arching. Squeezing the buttocks reduces the stress on the lower back.

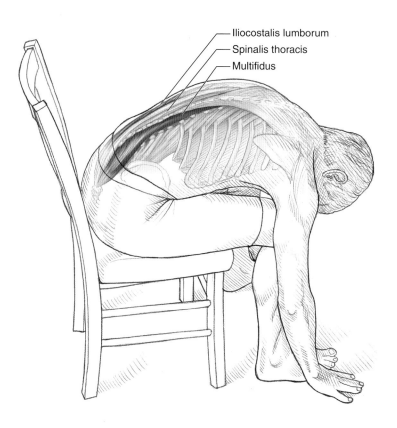

Iliocostalis lumborum
Spinalis thoracis
Multifidus

Execution

1. Sit upright in a chair, with legs separated.
2. Slowly round the upper back and begin to lean forward.
3. Continue to bend at the waist and lower the head and abdomen between the legs and below the thighs.

Muscles Stretched

Most-stretched muscles: Iliocostalis lumborum, multifidus

Less-stretched muscles: Interspinales, rotatores, spinalis thoracis

Stretch Notes

When done with incorrect posture, simple daily tasks, such as housecleaning, gardening, lifting heavy objects, and exercising, can cause tightness in back muscles. Poor posture includes slouching in chairs, sitting in hunched-back positions, standing in nonupright positions, and keeping the knees straight when lifting. All these actions lead to tight muscles by either overworking or overstretching the back muscles. Two other common reasons for tight back muscles are conscious psychological stress and subconscious repressed emotions. Stress causes back muscles to tighten in a fight or flight response, thus overworking the muscles and depriving them of energy needed to support the spine. In the short term, back stretching exercises reduce these problems by reducing stress. In the long term, these exercises make the back muscles stronger and longer and thus reduce the possibilities of overworking and overstretching.

Remember that hyperflexion can injure the spinal cord. When doing this exercise, go slowly and do not let the back become straight. Also, the effect of the stretch is minimized if the buttocks rise up off the chair.

⟨ VARIATION ⟩

Seated Lower-Trunk Extensor Lateral Flexor Stretch

Angling the head toward one of the knees will increase the stretch on the lower-trunk extensors and partially stretch some of the lateral flexors. Sit upright in a chair, with legs separated. Slowly extend the upper back and lean forward. As you lean forward, continue to bend at the waist and lower your head and abdomen toward the right knee. Finally, slowly lower the head below the right knee. Repeat toward the left knee.

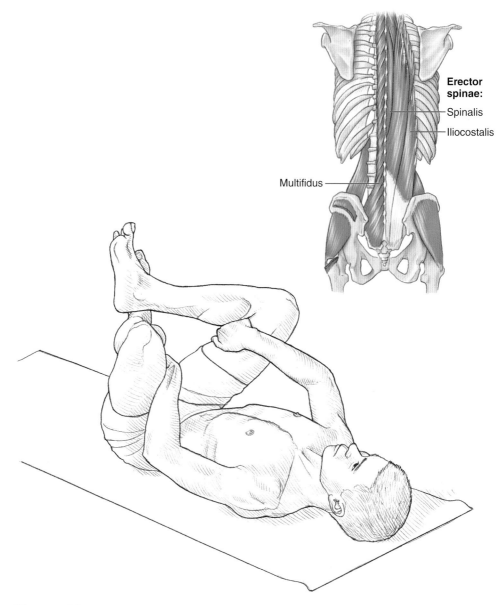

Erector spinae:

Spinalis

Iliocostalis

Multifidus

Execution

1. Lie on the back with the legs extended.
2. Flex the knees and hips, bringing the knees to the chest.
3. Cross the feet at the ankles, and separate the knees so they are at least shoulder-width apart.
4. Grasp the thighs inside of the knees, and pull the legs down to the chest.

Muscles Stretched

Most-stretched muscles: Iliocostalis lumborum, multifidus

Less-stretched muscles: Interspinales, rotatores, spinalis thoracis

Stretch Notes

Some people find that when they are performing the seated lower-trunk extensor stretch, they cannot lean forward slowly without contracting the back muscles. Keeping the muscles contracted while performing a stretch greatly reduces the effect of the stretch. Since the legs can weigh less than the trunk, these people may find it easier to perform this stretch while reclining. Also, since hyperflexion can injure the spinal cord, this stretch may be safer than the seated lower-trunk extensor stretch. When doing the reclining lower-trunk extensor stretch, it is easier to go slowly and not let the back straighten. By bringing the legs to the chest, you can easily raise the buttocks off the floor and prevent a straight back by allowing the spinal column to curl. Finally, do not try to bring the knees too far below the chest (do not try to touch the knees to the floor), as this could negate the safety benefits of this stretch.

LOWER TRUNK

BEGINNER LOWER-TRUNK LATERAL FLEXOR STRETCH

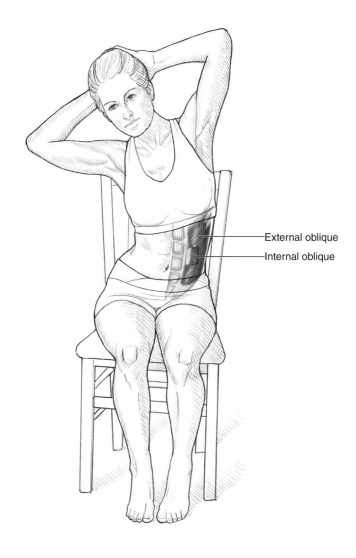

External oblique
Internal oblique

Execution

1. Sit upright in a chair.

2. Interlock the hands behind the head, with the elbows in a straight line across the shoulders.

3. While keeping both elbows back and in a straight line, laterally flex the waist and move the right elbow toward the right hip.

4. Repeat these steps for the opposite side.

Muscles Stretched

Most-stretched muscles: Left external oblique, left internal oblique, left rotatores

Less-stretched muscles: Left intertransversarii, left multifidus, left quadratus lumborum

Stretch Notes

Research has shown that the inability to do lateral flexion is a risk indicator for recurrent nonspecific low-back pain and injuries. Also, athletes who perform overhead actions for maximum distance or force, such as baseball players, football quarterbacks, and javelin throwers, need loose lateral flexors. They are also important for overhead hitting (e.g., racket-sport serves and smashes) and when reaching up as high as possible (e.g., rebounding a basketball or spiking a volleyball). Gymnasts, modern and ballet dancers, and divers need these muscles to be loose. In addition, tight lateral flexors can lead to a form of scoliosis. The quadratus lumborum's only action is lateral flexion, and tightness in this muscle results in a loss of lateral stability of the spine, causing the spine to curve to the left or right.

Flexing or extending at the waist will reduce this stretch's effectiveness. Also, keep the buttocks and thighs in complete contact with the chair. The closer the elbow gets to the floor, the harder it will be to remain seated in the chair. Wrapping the lower legs and feet around the chair legs will help keep the buttocks and thighs in contact with the seat.

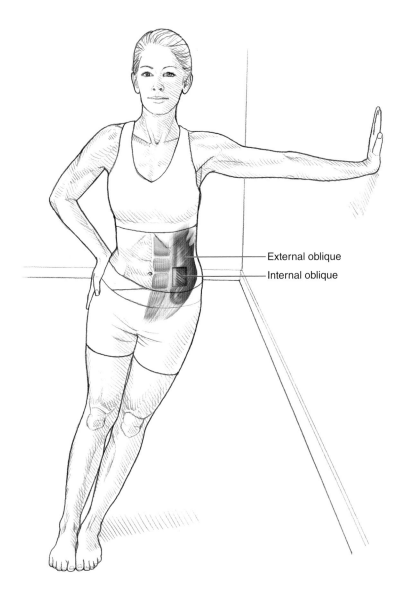

External oblique

Internal oblique

Execution

1. Stand upright with the feet together and the left side of the body facing a wall about an arm's length away.
2. Place the palm of the left hand on the wall at shoulder height. Place the heel of the right hand at the hip joint.
3. While keeping the legs straight, contract the buttocks and slightly rotate the hips in toward the wall.
4. Use the right hand to push the right hip toward the wall.
5. Repeat these steps for the opposite side.

Muscles Stretched

Most-stretched muscles: Left external oblique, left internal oblique, left rotatores

Less-stretched muscles: Left intertransversarii, left multifidus, left quadratus lumborum

Stretch Notes

Many sports rely on lateral trunk flexion. Since many of these activities stress one side of the body more than the other, it is easy for the two sides of the body to become unbalanced. The active side can become tight from being overworked. If the nonworking side goes unused for extended periods, the muscles can become short. Unbalanced body sides also can result from heavy lifting, especially if one side is substantially stronger, or from participation in activities such as martial arts and football in which the body receives heavy blows. This exercise is better suited than the basic lower-trunk lateral flexor stretch for restoring flexibility because the person is in a standing position similar to the detailed sports activities.

It is very easy to lose balance while doing this exercise, so stand on a nonskid surface. Keep the left arm straight, but do not lock the elbow. You can increase the amount of stretch by moving the feet farther from the wall, by resting the left forearm instead of the hand on the wall, or both.

Execution

1. Stand upright with legs 2 to 3 feet apart (60 to 90 cm), with the right foot about 1 foot (30 cm) ahead of the left foot.

2. Place both hands near the right hip.

3. Slowly arch the back, contracting the buttocks and pushing the hips forward.

4. As you continue to arch the back, rotate the trunk to the left and drop the head back toward the right side.

5. Slide the hands past the right buttock and down the right leg.

6. Repeat these steps for the opposite side.

Muscles Stretched

Most-stretched muscles: Rectus abdominis, left external oblique, left internal oblique

Less-stretched muscles: Left quadratus lumborum, left psoas major, left iliacus, left rotatores, left intertransversarii

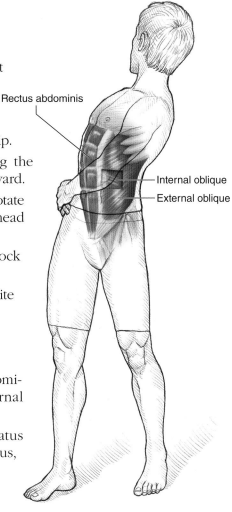

Rectus abdominis

Internal oblique

External oblique

Stretch Notes

This exercise is potentially dangerous, especially for people with a swayed back or weak abdominal muscles. This exercise can worsen a swayed back and cause excessive squeezing of the spinal discs, jammed spinal joints, and pinched spinal nerves emerging from the lumbar vertebrae. This stretch is recommended only for those who are very stiff and do not have a swayed back. Also, you should use this exercise only when the other lower-back flexor stretches do not provide any improvement. When doing this stretch, do minimal arching and make sure you squeeze the buttocks during the arching. Squeezing the buttocks reduces the stress on the lower back. Finally, it is very easy to lose balance while doing this exercise, so take extra care.

chapter 5

HIPS

The pelvic and the femur bones form the skeletal structure in the hip region of the body. The head of the femur bone fits into the acetabular fossa, a socket on the pelvis, to form the hip joint. This ball-and-socket joint allows the widest range of motion in the body. The movements of this joint include flexion, extension, abduction, adduction, and internal and external rotation of the hip. Surrounding the hip joints are several large and strong muscle groups, making possible the major movements of the lower extremities that are necessary for our daily activities.

Multiple muscles as well as several important ligaments surrounding the hip joint provide strong support. The ligamentum teres ligament connects the head of the femur and the acetabular notch of the pelvis to keep them together. The iliofemoral, ischiofemoral, and pubofemoral ligaments give extra support so that the head of the femur stays in the acetabular fossa in a firm, snug, and tight formation during all daily activities. The acetabular labrum runs along the rim of the acetabular fossa to deepen the hip cavity, thus giving additional support to the hip joint. All these structures combine to protect the hip joint and make it quite strong and able to withstand the demands of constant muscular movements.

All but two of the muscles of the hip (figure 5.1) run between the pelvic bones and the thigh bone (femur). The two exceptions are the psoas major and piriformis, which run between the lower vertebral column and the femur. The muscles that move the hip joint are some of the largest muscles (adductor magnus and gluteus maximus) in the body as well as some of the smallest (gemellus superior and inferior). The anterior (front) muscles—the psoas major, iliacus, rectus femoris, and sartorius—flex the hip and are used during walking to swing the leg forward. The posterior (back) muscles—the gluteus maximus, biceps femoris, semimembranosus, and semitendinosus—provide the backward swing for walking. A group of large muscles (adductor brevis, adductor magnus, adductor longus, gracilis, and pectineus) on the medial (inside) thigh keep the legs centered under the body. A group of small muscles (gluteus medius, gluteus minimus, piriformis, gemellus superior, gemellus inferior, obturator internus, obturator externus, quadratus femoris, and tensor fasciae latae) on the lateral (outside) thigh splay the legs to the side. Another group that makes up more than 75 percent of the hip muscles is the external hip rotators, consisting of the gluteus maximus, gluteus medius, gluteus minimus, piriformis, gemellus superior, obturator internus, gemellus inferior, obturator externus, quadratus femoris, psoas major, iliacus, rectus femoris, sartorius, adductor brevis, adductor magnus, adductor longus, and pectineus.

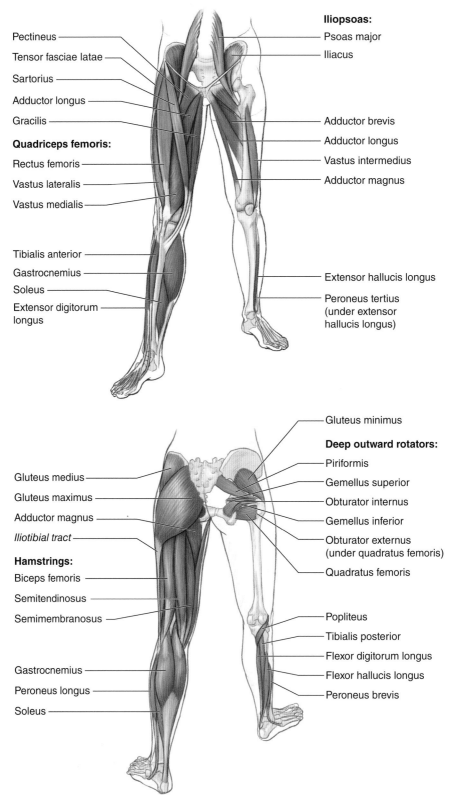

Pectineus

Tensor fasciae latae

Sartorius

Adductor longus

Gracilis

Quadriceps femoris:

Rectus femoris

Vastus lateralis

Vastus medialis

Tibialis anterior

Gastrocnemius

Soleus

Extensor digitorum longus

Iliopsoas:

Psoas major

Iliacus

Adductor brevis

Adductor longus

Vastus intermedius

Adductor magnus

Extensor hallucis longus

Peroneus tertius (under extensor hallucis longus)

a

Gluteus minimus

Deep outward rotators:

Piriformis

Gemellus superior

Obturator internus

Gemellus inferior

Obturator externus (under quadratus femoris)

Quadratus femoris

Gluteus medius

Gluteus maximus

Adductor magnus

Iliotibial tract

Hamstrings:

Biceps femoris

Semitendinosus

Semimembranosus

Gastrocnemius

Peroneus longus

Soleus

Popliteus

Tibialis posterior

Flexor digitorum longus

Flexor hallucis longus

Peroneus brevis

b

Figure 5.1   Muscles of the lower extremities: *(a)* anterior; *(b)* posterior.

The range of motion, or the degree of freedom to move the hip, depends on several factors, including bony structure; muscle strength; stiffness of muscle tissue, tendons, and ligaments; and anatomical restrictions. For hip flexion, the range of motion is limited by hip flexor strength, stiffness of the hamstring muscles, and contact of the leg with the abdomen. Extension is influenced by hip extensor strength and the stiffness of both the hip flexors and the ligaments surrounding the ball-and-socket joint. Hip abduction is limited not only by the strength and stiffness of the adductors but also by the stiffness of the pubofemoral and iliofemoral ligaments and bony contact of the femoral neck and acetabular rim. On the other hand, hip adduction is restricted by the strength of the adductors and stiffness of the abductors as well as the stiffness of the iliofemoral and capitate ligaments. Besides muscle strength of the agonist muscles and stiffness of the antagonist muscles, internal rotation movement is restrained by the iliofemoral and ischiofemoral ligaments, while external rotation is restrained by tension in the iliofemoral ligament.

Flexibility has more to do with overall body function than previously thought. For instance, diminished flexibility is one indicator of an aging body. Decreased physical activity also results in decreased flexibility. As people age and decrease their physical activity, they must keep stretching muscle groups in order to maintain mobility and range of motion in the joints. The hip region is located in the middle of the body, so problems in this area tend to radiate and affect many other parts of the body. You can reduce and even prevent many hip problems by paying more attention to strength and joint flexibility.

For instance, pain in the hip or buttocks area is often associated with poor hip flexibility. This is especially true after running or hiking along steep inclines or declines or even on slanted surfaces. Hip pain that occurs one or two days after activity is due to extensive use of the hip external rotator muscles and is caused by damage to both the muscle and the connective tissues in and around the muscle. Unfortunately, the hip external rotator muscles are small and usually weak and as such are not strengthened during typical strength-training activities. Therefore, stretching these muscles before and after the activity may help decrease this soreness and increase their strength. In addition, the hip external rotator muscles are the least-stretched muscles of the lower body, probably because these muscle groups are also the most difficult to stretch. We all tend to ignore those places in the body where we often find the most problems. On the bright side, it is easy to concentrate more on stretching those stiff and sore muscle groups.

The hip stretches in this book are grouped according to which muscle groups are being stretched. In addition, they are listed and described in order from the easiest to the most difficult. Stretches for the hip flexor muscles are explained first, followed by stretches for the hip extensors, hip adductors, and hip external rotators, in that order, from easiest to hardest in each category. Those who are new to a stretching program tend to be less flexible and should begin with the easiest level of stretches. Progression to a more difficult stretch in this program should be made when the participant feels confident he is able to advance to the next level. For detailed instructions, refer to the information on stretching programs in chapter 9.

It is also recommended that you explore the stretches in this book from different angles of pull. By slightly altering the position of the body parts, such as the hands or trunk, the pull of the muscle is changed. This approach is the best way to discover where the tightness and soreness in the specific muscles are located. Exploring different angles while stretching will also bring more versatility to your stretching program.

All the instructions and illustrations in this chapter are given for the right side of the body. Similar but opposite procedures are to be used for the left side of the body. The stretches in this chapter are excellent overall stretches; however, not all of these stretches may be completely suited to each person's needs. As a rule, to effectively stretch specific muscles, the stretch must involve one or more movements in the opposite direction of the desired muscle's movements. For example, if you want to stretch the right adductor magnus, perform a movement that involves extension, internal rotation, and abduction of the right leg. When a muscle has a high level of stiffness, use fewer simultaneous opposite movements. For example, to stretch a very tight adductor magnus, start by doing only hip abduction. As a muscle becomes loose, you can incorporate more simultaneous opposite movements.

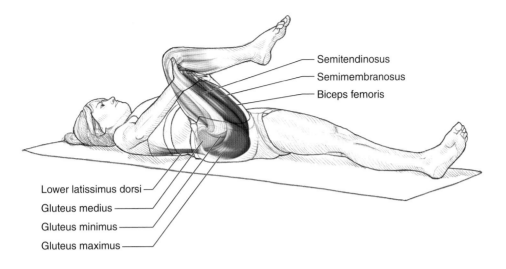

Semitendinosus
Semimembranosus
Biceps femoris

Lower latissimus dorsi
Gluteus medius
Gluteus minimus
Gluteus maximus

Execution

1. Lie on your back on a comfortable surface.
2. Bend the right knee, and bring it toward the chest.
3. While keeping the left leg flat, grasp the right knee with both hands, and pull it down toward the chest as far as possible.
4. Repeat this stretch for the opposite leg.

Muscles Stretched

Most-stretched muscles: Right gluteus maximus, right erector spinae, right lower latissimus dorsi, right semitendinosus, right semimembranosus, right biceps femoris

Less-stretched muscles: Right gluteus medius, right gluteus minimus

Stretch Notes

This is another helpful and effective stretch for people who suffer from lower-back and pelvic or hip pain. Pain in the pelvic region is often a result of muscular soreness, and when muscles are sore, they often feel stiff as well. A person with this condition has a tendency to limit the range of motion of the affected muscles in order to avoid pain. Therefore, normal daily activities can be significantly affected depending on the severity of the pain. Rather than avoiding movement, a person suffering from this condition should specifically try to move and stretch the injured muscles. Performing the hip and back extensor stretch will provide increased flexibility and strength to these muscle groups, which in turn will help lessen the likelihood (or severity) of future injury.

For warm-up purposes, it is recommended that you use both legs simultaneously at first. Once warmed up, bring one knee up to the chest at a time. In addition, pulling the knee up toward the armpit will maximize the effectiveness of this stretch.

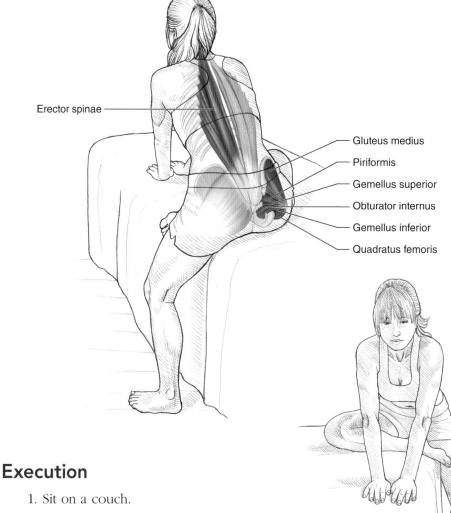

Erector spinae

Gluteus medius
Piriformis
Gemellus superior
Obturator internus
Gemellus inferior
Quadratus femoris

HIPS

Execution

1. Sit on a couch.

2. Rotate the right leg at the hip, and pull the right foot in to rest flat against the left inner thigh, as close as possible to the pelvic area. The outside of the lower right leg should rest as flat as possible on the surface of the couch.

3. Bend the trunk over toward the right (bent) knee as far as possible until you start to feel a slight stretch (light pain). Keep the left knee down, if possible, as you bend over.

4. As you bend over, reach out with your arms over the right foot.

5. Repeat this stretch for the opposite leg.

Muscles Stretched

Most-stretched muscles on right side: Gluteus maximus, gluteus medius, gluteus minimus, piriformis, gemellus superior, gemellus inferior, obturator externus, obturator internus, quadratus femoris

Most-stretched muscles on left side: Erector spinae, lower latissimus dorsi

Stretch Notes

This stretch is a lowest-stress version of the hip external rotator stretches and as such is the best stretch to use at first. The small hip external rotator muscles are located on the outer back side of the hip, underneath the gluteus maximus muscle. If you feel some minor tightness or soreness here, especially after walking, running activities, or climbing, use this low-intensity stretch to relieve the stress put on these muscles during these activities. You use these muscles whenever the hip rotates in an outward direction such as in walking and running. If the external rotator muscles are not strong or flexible enough, they can become sore and tight very easily.

This particular stretch can be easily done while sitting on a couch or bed, and it is one of the easiest stretches to execute for these muscles groups. Doing this stretching exercise in a sitting position with the right leg up flat, bent 90 degree or less, on the couch surface and the left leg hanging down is a relaxing position. If you are less flexible or a beginner to a stretching program, it would perhaps be better to start this stretch with your right knee in a less bent position (more than a 90-degree angle) and then gradually work on bending the knee more as your flexibility improves. Remember to bend the trunk in a forward direction from the hip. It is also beneficial to keep the back straight; do not curl or hunch the back while performing the stretch.

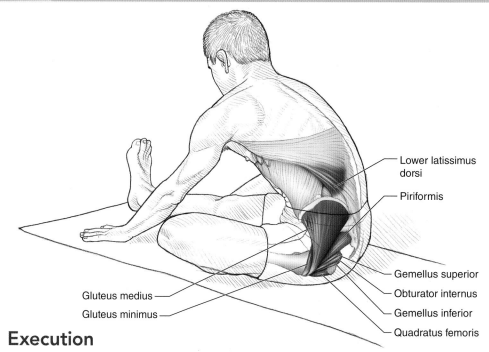

Lower latissimus dorsi

Piriformis

Gemellus superior

Obturator internus

Gemellus inferior

Quadratus femoris

Gluteus medius

Gluteus minimus

Execution

1. Sit with the right leg extended straight out in front. Bend the left knee and place the left foot flat against the right inner thigh, as close as possible to the pelvic area. Place the hands on the floor next to the thighs.

2. Keeping the trunk straight, bend the trunk forward from the hip joint over toward the right (straight) knee as far as possible until you start feeling a slight stretch (light pain). Keep the right knee down on the floor if possible as you bend over. Reach out with your arms toward the right foot.

3. Repeat this stretch for the opposite leg.

Muscles Stretched

Most-stretched muscles on left side: Gluteus medius, gluteus minimus, piriformis, gemellus superior, gemellus inferior, obturator externus, obturator internus, quadratus femoris, erector spinae, lower latissimus dorsi

Most-stretched muscles on right side: Semitendinosus, semimembranosus, biceps femoris, gluteus maximus, gastrocnemius

Less-stretched muscles on the right side: Soleus, plantaris

Stretch Notes

The hip external rotator muscles are commonly neglected in stretching routines. Overuse of these muscles in activities such as basketball, soccer, and hockey can lead to soreness, tightness, or even injuries to this area. In addition, poor flexibility usually leads to lower-quality performance. Participants do a lot of stepping sideways, using a lot of these muscles whenever the hip rotates outwardly. Utilizing this stretch regularly will build flexibility and strength.

Intermediate Seated Hip Extensor and External Rotator Stretch

Bending the trunk toward the left knee instead of the right knee reduces the stretch of the most-stretched muscles on the left side of the body and increases the stretch of the most-stretched muscles on the right side of the body. Sit with the right leg extended straight out in front. Bend the left knee and place the left foot flat against the right inner thigh, as close as possible to the pelvic area. Bend the trunk over toward the left (bent) knee as far as possible until you start feeling a slight stretch (light pain). Repeat on the opposite leg.

Semitendinosus
Semimembranosus
Gastrocnemius

Intermediate Seated Hip External Rotator, Extensor, Knee Flexor, and Plantar Flexor Stretch

Modify the intermediate seated hip external rotator and extensor stretch to include the soleus, popliteus, flexor digitorum longus, flexor hallucis longus, posterior tibialis, gastrocnemius, and plantaris muscles of the lower leg as a combo stretch. Sit with the right leg extended straight out in front. Flex the left knee and place the left foot flat against the right inner thigh, as close as possible to the pelvic area. Bend the trunk over toward the right (straight) knee as far as possible until you start feeling a slight stretch (light pain). As you bend forward, reach with the right arm, grasp the right foot, and pull the toes slowly toward the knee (dorsiflexed position).

Gastrocnemius
Soleus
Plantaris
Flexor digitorum longus
Tibialis posterior
Flexor hallucis longus

HIPS

ADVANCED STANDING HIP EXTERNAL ROTATOR STRETCH

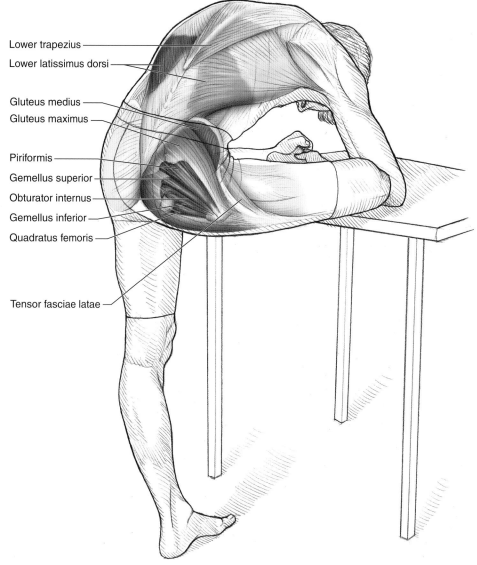

Lower trapezius

Lower latissimus dorsi

Gluteus medius

Gluteus maximus

Piriformis

Gemellus superior

Obturator internus

Gemellus inferior

Quadratus femoris

Tensor fasciae latae

Execution

1. Stand upright on the left leg, with the knee straight. Face a support surface such as a table, the edge of a couch, or a beam that is even with or just a little below the hips.

2. Bend the right leg at the hip at about a 90-degree angle and rest it on the support surface. The outside of the lower right leg rests as flat as possible on the surface. You can place a towel or pillow under the foot and lower right leg for cushioning.

3. Lower the trunk as far as possible toward the right foot, keeping the right knee as flat as possible on the surface.
4. Repeat this stretch for the opposite leg.

Muscles Stretched

Most-stretched muscles: Right gluteus maximus, right gluteus medius, right gluteus minimus, right piriformis, right gemellus superior, right gemellus inferior, right obturator externus, right obturator internus, right quadratus femoris, lower left erector spinae, lower left latissimus dorsi

Less-stretched muscles: Right tensor fasciae latae, right lower latissimus dorsi, lower right trapezius

Stretch Notes

It is not uncommon to encounter periodic extensive soreness or tightness in the hip area as a result of certain types of exercise movements. Often this is due to extensive use of the hip external rotator muscles in activities such as ice skating, in-line skating, or a skating style of cross-country skiing. These muscles are located in the deep tissue of the hip just underneath the gluteus maximus muscle.

This is a more advanced stretch than the previous stretches in this chapter. It is one of the best stretches for the hip external rotator muscles. When placing the bent right leg on the supporting surface, make sure the entire lower leg is resting on it. This helps put the lower leg in a position of minimal stress on the knee joint. In addition, placing extra cushioning underneath the bent leg will make this stretch more comfortable.

Be sure to lower the trunk forward from the hip joint as far as you can. Keep the trunk as a straight unit; do not let the back curve. Bending the trunk toward the right knee instead of the left knee reduces the stretch of the most-stretched muscles on the right side of the body and increases the stretch of the most-stretched muscles on the left side of the body.

In addition, slowly adding more height to the right leg placement (perhaps a couple of inches after every two to four weeks) makes this stretch even more demanding. Increasing the height of the table, bench, or other surface up to 1 foot (30 cm) above the hips will increase the stretch to the highest possible level for these muscle groups.

RECUMBENT HIP EXTERNAL ROTATOR AND EXTENSOR STRETCH

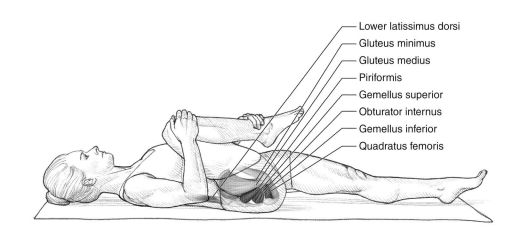

Lower latissimus dorsi
Gluteus minimus
Gluteus medius
Piriformis
Gemellus superior
Obturator internus
Gemellus inferior
Quadratus femoris

Execution

1. Lie on your back on a comfortable surface.

2. While outwardly rotating the right leg, bend the right knee and bring the right foot to the body's midline. The knee is aligned outside of the chest and is pointed laterally. While keeping the left leg flat, grasp the right knee with the right hand and the right ankle with the left hand. Pull the lower leg as a unit as far as possible toward the chest.

3. Repeat this stretch for the opposite leg.

Muscles Stretched

Most-stretched muscles: Right gluteus maximus, right piriformis, right gemellus superior, right gemellus inferior, right obturator externus, right obturator internus, right quadratus femoris, right lower latissimus dorsi, right erector spinae

Less-stretched muscles: Right gluteus medius, right gluteus minimus

Stretch Notes

This is another version of a low-stress stretch for the hip external rotator and hip extensor muscles. These particular muscles can become sore or tight after engaging in activities that are not usual to daily routines or when unusual stress is placed on them. For instance, playing an impromptu game of soccer with your kids or friends where sprinting, jumping, and making sudden changes of direction are required can easily result in uncomfortable or painful muscles later on. There are also times when soreness is experienced but it is difficult to recall what action or movement might have led to the aching muscles. In any case, when soreness or tightness is present, it is time to begin stretching the muscles affected. If you are new or relatively new to a stretching routine, this is a great stretch to begin with. As with many of the stretches in this book, it is easiest to begin a routine by sitting or lying down.

To maximize the effectiveness of this stretch, it is best to bring the ankle toward and over the head as far as possible. This will stretch the targeted muscles to the maximum level. Also, moving the ankle slightly to the right or left of the body will result in an additional pull on the multiple muscles of these hip rotators. Whenever you attempt any new or unaccustomed movement, as with the variations of this stretch, make sure to take safety precautions into consideration. In this case, put some extra support behind the left knee with the left hand or a towel. In a bent position, such as in this stretch, the knee is vulnerable to injury, especially during experimentation with new movements.

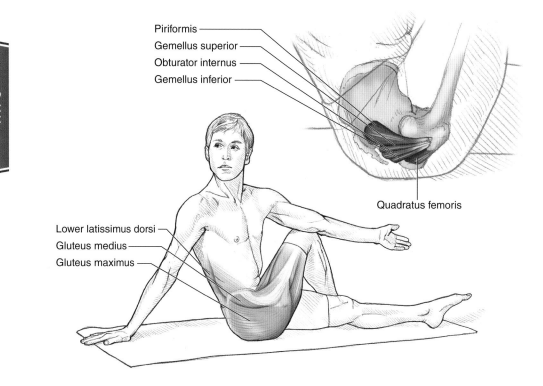

Piriformis
Gemellus superior
Obturator internus
Gemellus inferior

Quadratus femoris

Lower latissimus dorsi
Gluteus medius
Gluteus maximus

Execution

1. Sit on the floor with the left leg extended.

2. Bend the right leg, and place the right foot on the outside of the left knee.

3. Bend the left arm, and position the outside of the left elbow against the outside of the upraised right knee.

4. Brace the right arm against the floor near the right hip.

5. Push the left elbow against the right knee, twisting the trunk as far as possible to the right. Maintain enough pressure with the left elbow to keep the right knee in a stable position.

6. Repeat this stretch for the opposite leg.

Muscles Stretched

Most-stretched muscles on right side: Gluteus maximus, gluteus medius, gluteus minimus, piriformis, gemellus superior, gemellus inferior, obturator externus, obturator internus, quadratus femoris, lower latissimus dorsi, erector spinae

Less-stretched muscles on left side: Gluteus maximus, gluteus medius, erector spinae, lower latissimus dorsi

Stretch Notes

This low-intensity stretch is well suited for those who have lower-back and hip pain. Lower-back problems can be quite common among any adult population but tend to become more prevalent as one ages. Pain in this area might be attributed to a specific injury or might just accumulate with use of the back muscles over time. Another cause of lower-back pain and discomfort is weakness of the back and abdominal muscles or muscular imbalance between these two muscle groups. This condition also tends to radiate pain sensations to the pelvic area, possibly limiting one's ability to comfortably accomplish daily tasks. To help alleviate this pain and discomfort, it would be very beneficial to start performing this low-intensity stretch. Regular use of this stretch will strengthen this area and help reduce future reoccurrence of painful episodes.

While executing this stretch, try to keep the trunk upright. Do not arch the back or bend forward. Be careful to twist the trunk in a slow motion. This helps control the amount of stretch to the target muscles. Hold the position by bracing the right elbow against the left knee.

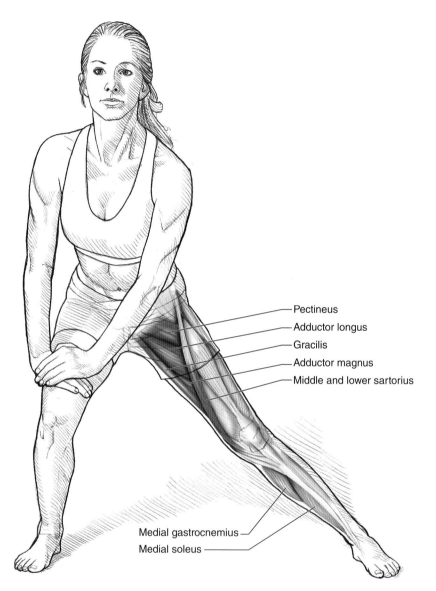

Pectineus
Adductor longus
Gracilis
Adductor magnus
Middle and lower sartorius

Medial gastrocnemius
Medial soleus

Execution

1. Stand upright with the legs more than shoulder-width apart and the left foot turned out.

2. Lower the hips to a half-squatting position, bending the right knee and sliding the left foot out to the left to keep the left knee straight.

3. Place the hands on top of the right knee for support and balance, or hold on to an object for balance.

4. Repeat this stretch for the opposite leg.

Muscles Stretched

Most-stretched muscles: Left gracilis, left adductor magnus, left adductor longus, left adductor brevis, left pectineus, middle and lower left sartorius, left semitendinosus, left semimembranosus

Less-stretched muscles: Medial left gastrocnemius, medical left soleus, left flexor digitorum longus

Stretch Notes

This is one of the easiest stretches for the inner thigh muscles. Most people do not use the inner thigh muscles to a great extent during normal daily activities. Consequently, these muscles tend to be weaker than other muscles in the thigh and hip areas and can become fatigued faster as a result. Participating in occasional activities such as walking or running on hilly terrain, climbing up and down stairs, or even playing a neighborhood basketball game with friends can sometimes lead to muscle twitching sensations, a sign of fatigue, in the inner thigh. If this occurs, it is recommended that the affected muscles be stretched for a couple of minutes to loosen them up. In most cases, the activity can then be resumed after stretching. It should be noted here that it is always beneficial to perform a series of light stretches before starting any type of exercise, sport, or strenuous activity. This decreases the possibility of injury or discomfort to any muscle group of the body.

While performing this stretch, keep the trunk as straight as possible. It is more comfortable to allow your weight to rest on the inside of the left foot. To increase the stretch, bend the trunk to the right, and press the right thigh down with both hands at the same time.

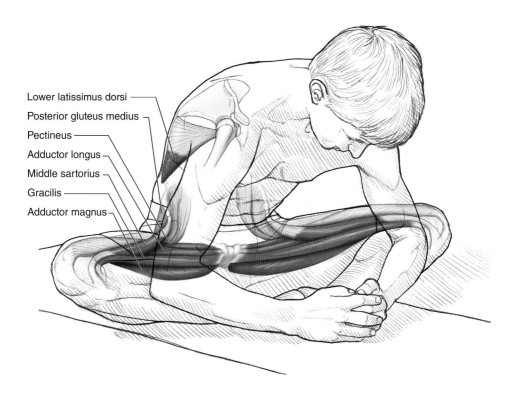

Lower latissimus dorsi
Posterior gluteus medius
Pectineus
Adductor longus
Middle sartorius
Gracilis
Adductor magnus

Execution

1. Sit on the floor in the lotus position—knees bent, feet together with the soles touching.

2. Bring the heels of the feet as close as possible to the buttocks. (Distance depends on the degree of your flexibility.)

3. Grasp the feet or just above the ankles, with the elbows spreading sideways and touching the legs just below the knees.

4. Bend the trunk over toward the feet, and press the lower part of the thighs and knees down with the elbows while stretching.

Muscles Stretched

Most-stretched muscles: Gracilis, adductor magnus, adductor longus, adductor brevis, pectineus, middle sartorius, lower erector spinae, lower latissimus dorsi

Less-stretched muscles: Gluteus maximus, posterior gluteus medius

Stretch Notes

The target muscles for this stretch—the adductor brevis, adductor longus, adductor magnus, gracilis, sartorius, and pectineus—are located on the medial (inner) side of the hip and thigh. These muscles are fairly large and are responsible for hip adduction (i.e., bringing the leg toward the midline of the body). Extensive use of hip adduction is typical in competitive or recreational activities such as ice skating, in-line skating, and the skating style of cross-country skiing. Most people engage in such activities on an occasional or seasonal basis. Unless training or conditioning is done as a regular routine, it is not uncommon for the more sporadic participant to encounter soreness or tightness after the activity. To prevent these symptoms from becoming more severe, it is recommended that these muscles be stretched before, during (if necessary), and after the activity.

The degree of stretch to the target muscles depends on the distance between the heels and the buttocks. The closer the heels are to the buttocks, the greater the stretch. In addition, the amount of stretch put on these adductor muscles can be controlled by the degree of pressure put on the lower part of the thighs and knees by the elbows. The stretch can be further intensified by grasping the feet and using them as a lever to pull the trunk forward. This technique not only targets the hip adductor muscles but acts as an effective stretch for the lower-back muscles as well. Placing the heels about 1 foot (30 cm) away from the buttocks increases the stretch on the gluteus maximus, gluteus medius, and erector spinae and places the greatest portion of the stretch on the origins of the adductor muscles.

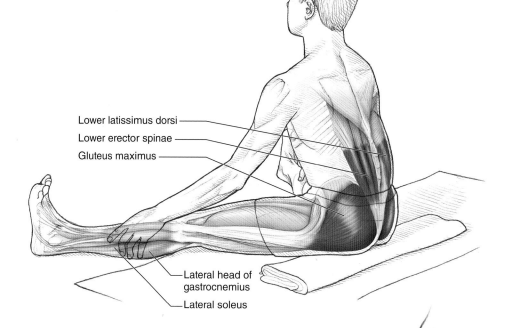

Lower latissimus dorsi

Lower erector spinae

Gluteus maximus

Lateral head of gastrocnemius

Lateral soleus

Execution

1. Sit comfortably on the floor with legs extended in a V position, feet as far apart from each other as possible.
2. Place the hands on the floor next to the thighs.
3. Keep both knees straight and as flat against the floor as possible.
4. Slide the hands forward along the legs, and bend the trunk over between the knees.

Hamstrings:

Biceps femoris

Semitendinosus

Semimembranosus

Muscles Stretched

Most-stretched muscles: Semitendinosus, semimembranosus, gracilis, adductor magnus, adductor longus, gluteus maximus, lower erector spinae, lower latissimus dorsi

Less-stretched muscles: Lateral soleus, lateral head of gastrocnemius, plantaris, biceps femoris

Stretch Notes

This is a more advanced stretch targeting the inside portion of the upper leg, the adductor muscles, as well as the inner backside of the thigh muscles, the semimembranosus and semitendinosus. In addition, it benefits the musculature of the lower back. Because of the nature of the position of this stretch, in which both legs are extended simultaneously, it is recommended for people who have already achieved a good amount of flexibility in this area of the body.

Keep both knees slightly bent while warming up. After the muscles are warmed up, you can move the knees into a straight position. To maximize the stretch, do not bend the knees, tilt the pelvis forward, or curve the back. Also, bend the trunk forward as a single unit, keeping it centered between the legs.

Changing the trunk position changes the nature of the stretch. For example, slowly moving the trunk in a position over the right knee puts more stretch emphasis on the right-side hip extensors, right lower-back muscles, and left-leg adductor muscles. Conversely, moving the trunk to a position over the left knee emphasizes the stretch in the left-side hip extensors, left lower-back muscles, and right-leg adductor muscles.

⟨ **VARIATION** ⟩

Seated Hip Adductor and Extensor Stretch With Toe Pull

By grasping the toes you can make this stretch more complex and thus increase its effectiveness by including additional muscles. You can stretch not only the calf, hamstrings, posterior hip, lower back, shoulder, and arm muscles but also the entire right and left sides of the body at the same time. The amount of stretch depends on how hard you pull the toes toward the knees and the tibia bone. Simply execute steps 1 to 3 of the Seated Hip Adductor and Hip Extensor Stretch, and then for step 4 simply grasp the toes of both feet and pull them toward your head.

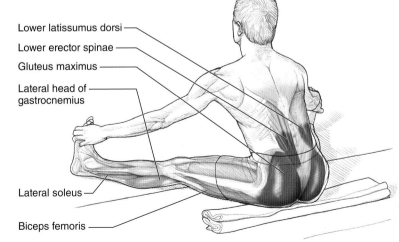

Lower latissumus dorsi

Lower erector spinae

Gluteus maximus

Lateral head of gastrocnemius

Lateral soleus

Biceps femoris

chapter 6

KNEES AND THIGHS

The skeletal structure of the upper leg and knee is made up of the tibia and fibula (lower leg) and the femur (upper leg). These long bones in the lower and upper regions of the leg form the major lever system that allows the body to use the muscles of this region in all locomotive movements.

The knee joint is the only major joint between the bones of the lower and upper leg. It is classified as a hinge joint, and it allows only two major movements, flexion and extension. The range of motion, or the degree of freedom to move this joint, depends on both the bone structure and the flexibility of the muscle tissue, tendons, and ligaments that surround this joint. Typically, the knee joint is rather limited in movement compared with some other joints in the body, but the combination of the knee and the hip joint allows us to perform a variety of complicated movements and can enhance various sports and leisure activities. The more flexible these muscles are, the more freedom of movement possible.

The knee is surrounded by a number of ligaments and tendons (figure 6.1) to bring more stability. In spite of these additional supportive structures, the knee is still quite vulnerable to a number of injuries. One of the most important ligaments around the knee is the patellar ligament. It extends from the patella to the upper front tibia. The tendons of the quadriceps muscles, located in the front thigh, blend with the patellar ligament, which attaches these muscles to the tibia. The medial collateral ligament supports the medial (inner) side of the knee, and the lateral (outer) side of the knee is supported by the lateral collateral ligament. The anterior and posterior cruciate ligaments help prevent anterior and posterior displacements of the femur on the tibia bone. These ligaments are located inside the knee and hold the tibia and femur bones together. The oblique popliteal and arcuate popliteal ligaments provide additional support to the lateral posterior (outer back) area of the knee.

In addition, the medial and lateral patellar retinacula also arise from the quadriceps tendon and contribute to anterior support of the knee. Finally, a meniscus sits on the plateau (top) of the tibia, which gives additional stability to the knee and cushions the bones during walking, running, and jumping. Wear and tear of these menisci bring pain most often to the medial (inner) side of the knee joint.

Most of the muscles that control the movements of the knee are found in the thigh. However a few calf muscles are also involved. Generally, the thigh muscles that move the knee are categorized into two groups. The four large anterior thigh muscles—rectus femoris, vastus intermedius, vastus lateralis, and vastus medialis—are collectively called the quadriceps muscles, and these are the major knee extensors. The large posterior thigh muscles—biceps femoris, semimembranosus, and semitendinosus—are collectively called the hamstring muscles, and these are the major knee flexors. The hamstrings are assisted in

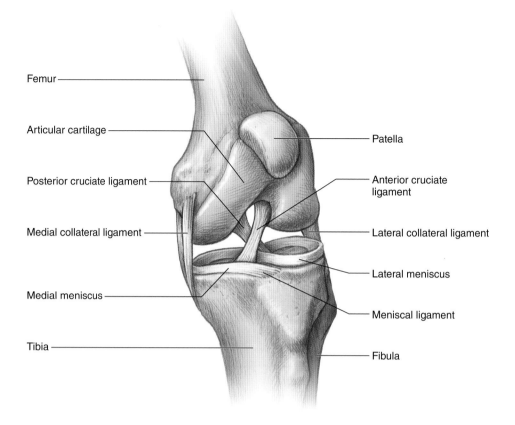

Femur

Articular cartilage

Posterior cruciate ligament

Medial collateral ligament

Medial meniscus

Tibia

Patella

Anterior cruciate ligament

Lateral collateral ligament

Lateral meniscus

Meniscal ligament

Fibula

Figure 6.1   Knee ligaments and tissue.

knee flexion by the gracilis and sartorius on the medial side of the thigh and the gastrocnemius, popliteus, and plantaris on the posterior side of the lower leg.

Flexion and extension are the two major movements of the knee. Most muscles in the body cross several joints, and thus many of these muscles are able to do several movements. Three of the quadriceps muscles, the vastus muscles, cross only one joint. This muscular arrangement allows these muscles to perform only knee extension. These three vastus muscles are strong extensors and sometimes may be sore and tight in front of the knee where the patella bone is located. Muscle tightness due to lack of stretching the quadriceps muscles is most often the cause of this problem. The knee extensors tend to exert less movement in walking, running, or jumping than the hamstring muscles. On the other hand, the hamstring muscles have two major movements—knee flexion and hip extension—and are active during any locomotive movement of the body. Thus, it appears that more total load is put on the hamstring muscles than on the quadriceps muscles. Because of this factor, the hamstring muscles tend to become more fatigued and sore than the quadriceps muscles during daily activities.

The muscles of the thigh that control the knee are important in all motor movements. Being much larger than the muscles of the calf and foot, the thigh muscles are better able to withstand muscular stress. Hence, muscular soreness occurs less often in these muscle groups. It is important, however, to have the right balance of strength and flexibility between the opposing muscle groups of the thigh. Most people have stronger but less flexible quadriceps muscles than hamstring muscles. People tend to stretch the hamstring muscles much more than the quadriceps muscles. This creates an imbalance between the two muscle groups. Chronic overstretching of the hamstrings without comparable stretching of the quadriceps can cause more harm than good. This is the reason hamstring muscles are sore more often than quadriceps muscles. Overstretching can also lead to chronic fatigue and a decrease in strength in the hamstring muscles. To correct this imbalance, you need to put more emphasis on quadriceps stretching and decrease the emphasis on hamstring stretching.

People often sit in one position for a long time, especially when in a car, behind a desk, or on an airplane. Thus, it is not surprising that after sitting for hours, people feel the need to get up and stretch their muscles. When people do stand up after long periods of sitting, they typically find that their joints and muscles have become temporarily stiff. Most often you feel more stiffness in the knee joint, and getting up from a long sitting position could be a rather painful experience. Because of this, it is recommended to get up often during those long sitting hours and move around. Stretching these muscles is a natural remedy. Many people have found that stretching and moving the leg muscles provides relief from muscular and joint tension and pain. Since muscular soreness and tension are common in the thigh muscles, both temporary and lasting relief can be obtained from a regular daily stretching routine. This routine needs to be a consistent part of a fitness program.

The knee and thigh stretches in this book are grouped according to which muscle groups are being stretched. In addition, they are listed and described in order from the easiest to the most difficult. Stretches for the hamstrings are explained first, followed by stretches for the quadriceps, from easiest to hardest. Those who are new to a stretching program tend to be less flexible and should begin with the easiest level of stretches. Progression to a more difficult stretch in this program should be made when the participant feels confident she is able to advance to the next level. For detailed instructions, refer to the information on stretching programs in chapter 9.

It is also recommended that the stretches in this book be explored from different angles of pull. Slightly altering the position of the body parts, such as the hands or trunk, changes the pull of the muscle. This approach is the best way to discover where the tightness and soreness in the specific muscles are located. Exploring different angles while stretching will also bring more versatility to your stretching program.

All the instructions and illustrations are given for the right side of the body. Similar but opposite procedures are to be used for the left side. The stretches in this chapter are excellent overall stretches; however, not all of these stretches may be completely suited to each person's needs. As a rule, to effectively stretch specific muscles, the stretch must involve one or more movements in the opposite direction of the desired muscle's movements. For example, if you want to stretch the right biceps femoris, perform a movement that involves extension and external rotation of the right leg. When a muscle has a high level of stiffness, use fewer simultaneous opposite movements. For example, to stretch a very tight biceps femoris, start by doing only knee extension. As a muscle becomes loose, you can incorporate more simultaneous opposite movements.

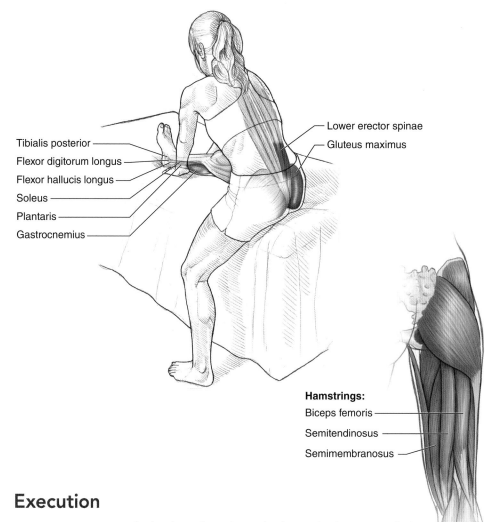

Tibialis posterior

Flexor digitorum longus

Flexor hallucis longus

Soleus

Plantaris

Gastrocnemius

Lower erector spinae

Gluteus maximus

Hamstrings:

Biceps femoris

Semitendinosus

Semimembranosus

Execution

1. Sit on a couch, bed, or bench with the right leg extended on the surface.

2. Rest the left foot on the floor, or let it hang down in a relaxed manner.

3. Place the hands on the couch, bed, or bench next to the right thigh or knee.

4. Bend at the waist and lower the head toward the right knee, keeping the back of the right knee comfortably on the couch, bed, or bench as much as possible.

5. While bending forward, slide the hands toward the right foot, keeping them alongside the lower leg.

6. Repeat this stretch for the opposite leg.

Muscles Stretched

Most-stretched muscles: Right semitendinosus, right semimembranosus, right biceps femoris, right gluteus maximus, right gastrocnemius, right lower erector spinae

Less-stretched muscles: Right soleus, right plantaris, right popliteus, right flexor digitorum longus, right flexor hallucis longus, right posterior tibialis

Stretch Notes

Tight knee flexors or hamstring muscles affect posture and the way the body moves during exercise. When these muscles are tight, the pelvis and hips are pulled out of their natural alignment, resulting in a flattened back and loss of the natural curve. A flatter lower back puts increased pressure on the sciatic nerve that runs down the legs and can cause muscles to tighten more. When the muscles are tight, they are also short, and short knee flexors increase the strain on the lower-trunk extensor muscles, especially when you bend forward at the waist. This added strain then injures the lower-trunk extensor muscles and is one of the most common causes of a sore lower back. Also, a lack of flexibility in the knee flexors makes these muscles more injury prone when a person suddenly increases movement speed or experiences greater workloads.

There are many reasons why an inactive person may have short knee flexor muscles. First, you can be born with naturally short hamstrings. Second, the hamstrings can become short if you sit for long periods. Regardless of the reason, your hamstrings can become longer if you perform regular stretching exercises.

Stretching the knee flexors one leg at a time reduces the stress on the legs and back. The knee flexor stretch can be performed on a soft couch or other soft surfaces and can be done at any time—while sitting on the couch watching TV or just relaxing after a long day's work. Doing this stretching exercise from a sitting position with one leg up on the couch surface and the other leg hanging down allows you to concentrate solely on stretching these muscles and allowing the other muscles in the body to relax. If you are not flexible or are a beginner to a stretching program, it would perhaps be better to start this stretch with your right knee slightly bent and then gradually work on straightening the knee as your flexibility improves. If you want to maximize the stretch of these muscles, start working with a straight knee position. While executing this stretch, try to avoid tilting the pelvis forward or curving the back. It is also beneficial to bend the trunk forward as a single unit, keeping it centered above or next to the side of the right thigh.

INTERMEDIATE STANDING KNEE FLEXOR STRETCH

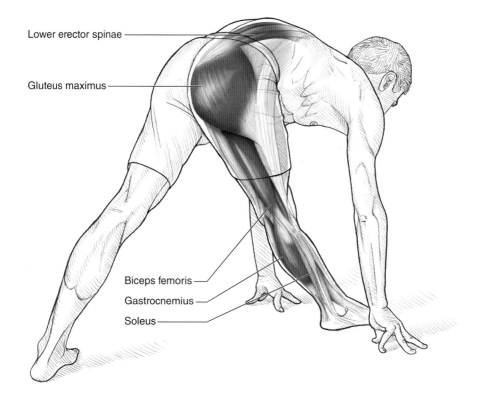

Lower erector spinae

Gluteus maximus

Biceps femoris

Gastrocnemius

Soleus

Execution

1. Stand upright with the right heel a comfortable distance ahead of the left toes.
2. Keeping the right knee straight and the left knee slightly bent, bend the trunk over toward the right knee.
3. Reach the hands toward the right foot.
4. Repeat this stretch for the opposite leg.

Muscles Stretched

Most-stretched muscles: Right semitendinosus, right semimembranosus, right biceps femoris, right gluteus maximus, right gastrocnemius, lower right erector spinae

Less-stretched muscles: Right soleus, right plantaris, right popliteus, right flexor digitorum longus, right flexor hallucis longus, right tibialis posterior

Stretch Notes

When you start participating in a sport and do not stretch properly, you are more likely to have your hamstrings tighten up. Tight hamstrings are common among both distance runners and sprinters who have significantly increased their speed, the distance run, or the amount of uphill climb. Tightness in the muscles can ease away during exercise as the muscles get warmer, but after stopping it can return. Also, tightness is often an indicator of minor or major muscle strains, a common occurrence mainly felt postexercise. In addition, muscle strength imbalances, in which the knee extensors are stronger or the gluteal muscles are weaker than the hamstrings, will also cause tightness. Thus, it is especially important to stretch properly after exercise because this is when the muscles are warm and more receptive to stretching.

This is the most commonly used stretch for the hamstring muscles. It can easily be done at any time whenever you feel the need to stretch your hamstrings. After any type of fitness activity, minor aches and tightness in the hamstring muscles are possible. It is not unusual to have such discomfort after almost any exercise session. This is the optimal time to do some light stretches for these muscles. In most cases, this stretch will relieve those uncomfortable symptoms, and you will be able to continue on to your other daily routines without any concern about your muscle condition.

For the best results in this stretch, try to keep the right knee straight and bend the torso directly from the hip. It is also important to keep the back as straight as possible when executing this stretch. Turning the right foot out slightly and bending the head and trunk more toward the medial (inner) side of the right knee will increase the stretch of the biceps femoris, which is located on the back outer side of the thigh. On the other hand, turning the right foot in slightly and bending the head and trunk more toward the lateral (outer) side of the knee will increase the stretch of the semitendinosus and semimembranosus muscles located on the back inner side of the thigh.

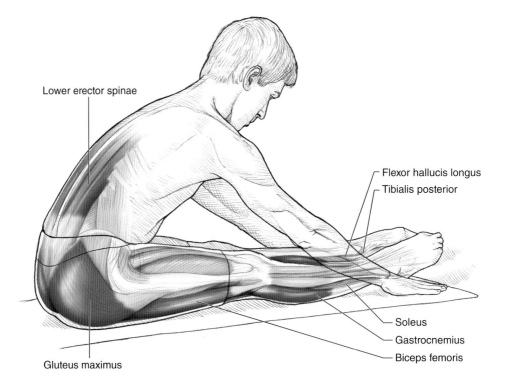

Lower erector spinae

Flexor hallucis longus

Tibialis posterior

Soleus

Gastrocnemius

Biceps femoris

Gluteus maximus

Execution

1. Sit on the floor, rug, or exercise mat with both legs extended and the insides of the ankles as close together as possible.
2. Keep the feet relaxed in a natural position.
3. Place the hands on the floor next to the thighs.
4. Bend at the waist and lower the head toward the legs. If possible, keep the back of the knees on the floor.
5. While bending forward, slide the hands toward the feet, and keep them alongside the legs.

Muscles Stretched

Most-stretched muscles: Semitendinosus, semimembranosus, biceps femoris, gluteus maximus, gastrocnemius, lower erector spinae

Less-stretched muscles: Soleus, plantaris, popliteus, flexor digitorum longus, flexor hallucis longus, posterior tibialis

Stretch Notes

When the hamstrings are tight, the pelvis and hips are pulled out of their natural alignment, resulting in a flattened back and loss of the natural curve. A flatter lower back puts increased pressure on the sciatic nerve that runs down the legs and can cause muscles to tighten more. Tight muscles are also short muscles, and short knee flexors increase the strain on the lower-trunk extensor muscles, especially when you bend forward at the waist. This added strain then injures the lower-trunk extensor muscles, one of the most common causes of a sore lower back. A tight muscle can compress the blood vessels within the muscle, and the reduced blood flow can make the hamstrings and lower-back extensors tighter and more fatigued.

Although this stretch helps alleviate problems by increasing flexibility, it is not recommended until you have already increased hamstring flexibility. If this exercise is done when both sets of muscles are tight, you risk causing damage to the lower back. This is because the hamstrings are usually larger and stronger, and so the weaker link gives out first.

In this exercise, you are able to stretch both legs at the same time. For you to stretch better and improve your flexibility, try your best to keep the knees straight. It is also important to keep the back straight. When you bend the trunk forward, try to move it as a single unit, keeping it centered between your legs. Following these procedures allows you to stretch the hamstring muscles more effectively and brings you more enjoyable, quicker, and better results.

It is usually more comfortable to perform this stretch on a carpet, exercise mat, or other soft surface. Doing this stretching exercise in a sitting position allows you to relax the other muscles in your body. This stretch could easily be performed while you are just sitting around, watching TV, reading, or doing any sitting leisure activity. Because we all do a lot of sitting in a day, this stretch can be done at any time and repeated throughout the day. A concentrated effort to remind yourself to perform this stretch daily might be very helpful in accomplishing this goal.

⟨ VARIATION ⟩

Seated Knee, Ankle, Shoulder, and Back Stretch

Instead of leaving the hands alongside the legs, if you grasp the toes and pull them slowly toward the knees (dorsiflexed position), you add the calf muscles to this stretch. In addition, changing to this hand position stretches the back, shoulder, and arm muscles. Simply follow steps 1 to 4 as detailed previously. Once you are in the step 4 position, grasp the toes or balls of the feet and pull the feet toward the knees.

Soleus

Gastrocnemius

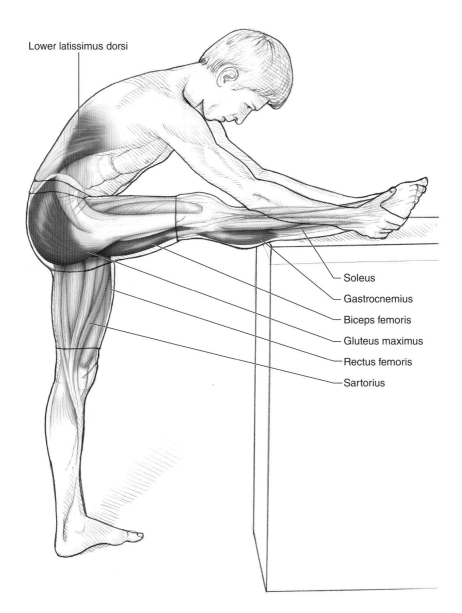

Lower latissimus dorsi

Soleus
Gastrocnemius
Biceps femoris
Gluteus maximus
Rectus femoris
Sartorius

Execution

1. Stand upright with your weight balanced on the left leg.
2. Flex the right hip and place the right leg, with the knee straight, on a table, bench, or other stable object that is higher than the height of the hips.
3. Bend at the waist, extend your arms over the lower right leg, and lower the head toward the right leg, keeping the right knee as straight as possible.
4. Keep the left knee straight and the left foot pointing in the same direction as the right leg.
5. Repeat this stretch for the opposite leg.

Muscles Stretched

Most-stretched muscles: Right gluteus maximus, right semitendinosus, right semimembranosus, right biceps femoris, right erector spinae, lower right latissimus dorsi, right gastrocnemius

Less-stretched muscles: Right soleus, right popliteus, right plantaris, right flexor digitorum longus, right flexor hallucis longus, right posterior tibialis, left sartorius, left rectus femoris

Stretch Notes

This is a more advanced stretch for those whose knee flexors are already more flexible than the average athlete. Be sure to select the right starting height for the table, bench, couch, or other stable object you will put your leg on. At the beginning of your stretching program, it is recommended to start at a lower height based on your state of flexibility and then periodically increase the height of the surface by several inches as your flexibility improves. Increasing the height of the surface by 1 to 2 feet (30 to 60 cm) above the hips as your flexibility improves will increase the stretch of these muscle groups. At this point, you will also start feeling a stretch to a portion of the front part of the left-leg muscle groups (sartorius; rectus femoris; vastus intermedius, lateralis, and medialis) as you increase the table to the highest possible height.

To maximize the stretch of the knee flexors, do not bend the knees, tilt the pelvis forward, or curve the back. In addition, bend the trunk straight forward as a single unit, keeping it centered over the right leg.

⟨ VARIATION ⟩

Raised-Leg Knee, Ankle, Shoulder, and Back Stretch

Grasping and pulling on the toes adds more muscles to the stretching process. This combination exercise stretches most of the posterior (back) body muscles at the same time. This naturally saves some time if your exercise time is limited. Follow steps 1 to 3. Once you are in the step 4 position, grasp the toes or the ball of the foot and pull the foot toward the knee.

Lower latissimus dorsi

Gluteus maximus

Gastrocnemius

Biceps femoris

Rectus femoris

Sartorius

RECUMBENT KNEE FLEXOR STRETCH

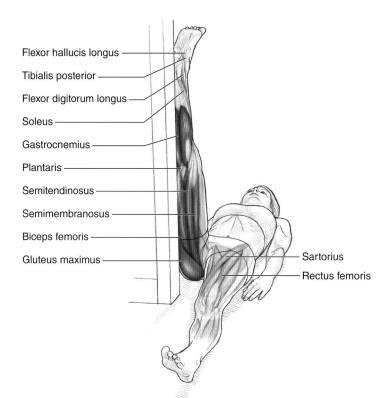

Flexor hallucis longus
Tibialis posterior
Flexor digitorum longus
Soleus
Gastrocnemius
Plantaris
Semitendinosus
Semimembranosus
Biceps femoris
Gluteus maximus

Sartorius
Rectus femoris

Execution

1. Lie flat on your back in a doorway, with the hips placed in front of the doorframe.
2. Raise the right leg and rest it on the doorframe. Keep the right knee straight and the left leg flat on the floor.
3. Place the hands palms down on either side of the buttocks.
4. Keeping the right leg straight, use the hands to slowly move the buttocks through the doorframe until you feel a stretch in the back of the leg.
5. Repeat this stretch for the opposite leg.

Muscles Stretched

Most-stretched muscles: Right gluteus maximus, right semitendinosus, right semimembranosus, right biceps femoris, right gastrocnemius

Less-stretched muscles: Right soleus, right popliteus, right plantaris, right flexor digitorum longus, right flexor hallucis longus, right posterior tibialis, left sartorius, left rectus femoris

Stretch Notes

When stretching the knee flexors, you must be careful of the lower back. If the lower-back extensor muscles are tight, they will limit the ability to perform most knee flexor stretches. As a result many people overstress the back. Also it is easy to tilt the pelvis forward or curve the back. Doing this can further harm the lower-back muscles. When you are lying down on your back, it is easier to maintain correct back positioning, and the floor provides additional back support. Thus, this exercise is the best knee flexor stretch to use when you have back problems.

Positioning your body in the right place for this stretch could take some extra time and effort, but once you are able to find the right position, it is an excellent stretch. To maximize the stretch of the knee flexors, do not bend the knees, tilt the pelvis forward, or round the back. Adjust the distance between the buttocks and the doorframe to increase or decrease the stretch. The closer the buttocks are to the doorframe, the greater the stretch. Once the buttocks cannot be positioned any closer to the doorframe, bending the right leg at the hip and moving the right leg toward the head can increase the stretch. It is also important to keep the left leg straight in front of you on the floor in order for you to get the maximal effect of this stretch. When reaching the maximal stretch limit, you will find that the quadriceps muscles on the left leg are getting stretched as well.

⟨ VARIATION ⟩

Recumbent Knee, Ankle, Shoulder, and Back Stretch

Using a towel to pull the toes downward increases the number of muscles being stretched. Follow steps 1 to 4. Once you feel the hamstrings being stretched, use the towel to pull the toes and foot down toward the floor. This combo exercise stretches most of the muscles on the back side of the body at the same time. In other words, you can work on multiple muscle groups, including the calf, hamstring, back, shoulder, and arm muscles, to a small degree. This naturally saves some time if your exercise time is limited.

KNEES · THIGHS

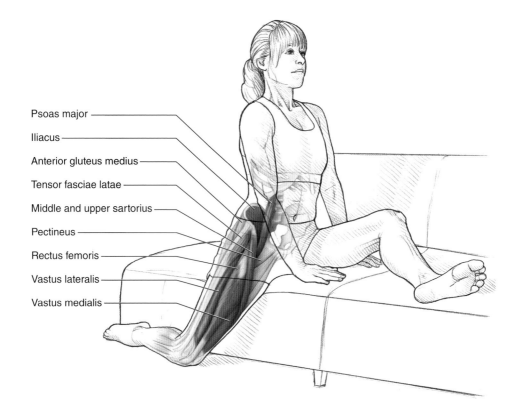

Psoas major
Iliacus
Anterior gluteus medius
Tensor fasciae latae
Middle and upper sartorius
Pectineus
Rectus femoris
Vastus lateralis
Vastus medialis

Execution

1. Sit upright on a couch or bed, with the left knee bent at less than a 90-degree angle in front of you. The lateral side of the left leg should be flat on the surface and the left hip on the edge of the couch or bed.

2. Balance the weight of your body over the left hip.

3. Extend the right leg behind the torso, and touch the floor with the right knee. The lower right leg lies on the floor.

4. Place the hands on the couch or bed to maintain balance.

5. Move the hips slowly forward, if needed, for more stretch.

6. Repeat this stretch for the opposite leg.

Muscles Stretched

Most-stretched muscles: Right vastus medialis, right vastus intermedius, right vastus lateralis, middle and upper right sartorius, right rectus femoris, right psoas major, right iliacus, right tensor fasciae latae

Less-stretched muscles: Right pectineus, anterior right gluteus medius

Stretch Notes

The knee extensors, the quadriceps, are used for common actions such as standing, sitting, walking, running, and jumping. Strains and injuries of the quadriceps muscles and tendons are common among athletes 15 to 30 years old who are engaged in explosive activities. On the other hand, for people engaged in daily living activities, the average age for injuries to these muscles is 65. Muscle strains and tears usually happen when a muscle is stretched beyond its limit, tearing the muscle fibers. They frequently occur near the point where the muscle joins the tendon. The four main causes of quadriceps injuries are muscle tightness, muscle imbalance, poor conditioning, and muscle fatigue. The ease of performing this beginning-level stretch will hopefully motivate you to stretch these muscles, especially since it can be done while reading, watching TV, or just relaxing.

This is a beginner's stretch for the quadriceps muscles. You can perform this stretch while sitting on the edge of a sofa or bed. The sitting position helps make the execution of this stretch more comfortable and relaxing. Place a pillow under the right knee for added comfort. Having the left leg in the bent position in front of you allows the stretch to be focused on the right leg's quadriceps muscles. Extend the right leg back from the torso.

Moving your hips slowly forward allows you to monitor the amount of stretch you put on the quadriceps muscles. This stretch can increase in intensity as needed or desired. Simply try arching the back slightly while moving the hips in the forward direction. After maximizing this level of stretching, start using the more advanced stretches found in this chapter.

KNEES · THIGHS

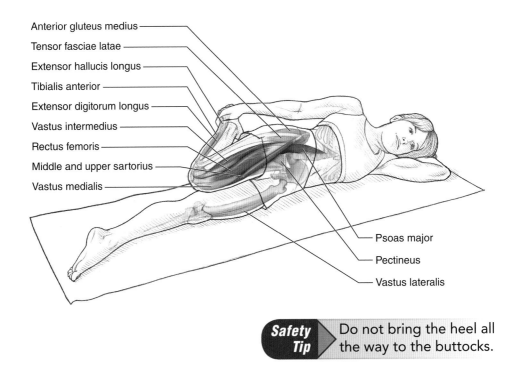

Anterior gluteus medius
Tensor fasciae latae
Extensor hallucis longus
Tibialis anterior
Extensor digitorum longus
Vastus intermedius
Rectus femoris
Middle and upper sartorius
Vastus medialis

Psoas major
Pectineus
Vastus lateralis

Safety Tip Do not bring the heel all the way to the buttocks.

Execution

1. Lie on the left side of the body.
2. Bend the right knee, and bring the right heel to within 4 to 6 inches (10 to 15 cm) of the buttocks.
3. Grasp the right ankle tightly, and pull the leg back close to your buttocks. However, do not bring the heel of the right ankle all the way to the buttocks.
4. At the same time, push the hip forward.
5. Repeat this stretch for the opposite leg.

Muscles Stretched

Most-stretched muscles: Right vastus intermedius, right rectus femoris, right psoas major, middle and upper right sartorius

Less-stretched muscles: Right vastus medialis, left vastus lateralis, right tensor fasciae latae, right pectineus, right iliacus, anterior right gluteus medius, right tibialis anterior, right extensor digitorum longus, right extensor hallucis longus

Stretch Notes

Injuries to the quadriceps muscles usually occur during an activity such as sprinting, jumping, or kicking, especially when the muscles are tight or unprepared for activity. This is yet another effective method of stretching the front thigh muscles. Although slightly more difficult than the beginner seated knee extensor stretch, this stretch still falls within the advanced beginner or intermediate category.

Because you perform this stretch while in a relaxed position, you have maximum control over the amount of stretch to the quadriceps muscles. In other words, this stretch allows you to concentrate solely on these thigh muscles while letting other muscles be as relaxed as possible.

Slowly pull the ankle in a more backward rather than upward direction while making sure the hips are also moving forward. Concentration should be greater on the forward hip movement than on the knee flexion (pulling the ankle toward the buttocks). As in any quadriceps muscle stretch, take extra care to prevent strain on the knee structure by overflexing the knee.

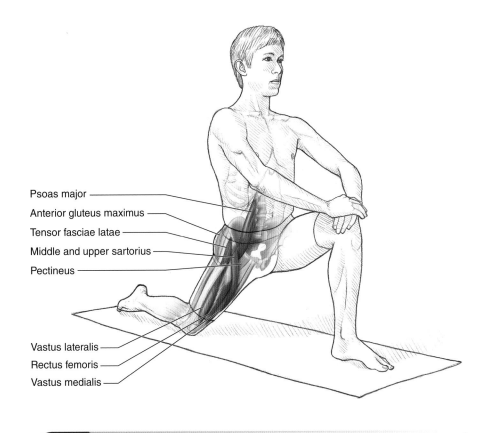

Psoas major

Anterior gluteus maximus

Tensor fasciae latae

Middle and upper sartorius

Pectineus

Vastus lateralis

Rectus femoris

Vastus medialis

Safety Tip ▶ Do not attempt this stretch until you have moved past the beginner and intermediate knee extensor stretches.

Execution

1. Step forward with the left leg, and bend the knee at about a 90-degree angle.

2. Keep the left knee positioned above the left ankle.

3. Extend the right leg behind the torso, and touch the floor with the right knee. The lower right leg lies on the floor.

4. Hold on to an object or place the hands on the left knee to maintain balance.

5. Move the hips forward, pushing the left knee in front of the left ankle and dorsiflexing that ankle.

6. Repeat this stretch for the opposite leg.

Muscles Stretched

Most-stretched muscles: Right vastus medialis, right vastus intermedius, right vastus lateralis, middle and upper right sartorius, right rectus femoris, right psoas major, right iliacus, right tensor fasciae latae

Less-stretched muscles: Right pectineus, anterior right gluteus maximus

Stretch Notes

The advanced kneeling knee extensor stretch is the quadriceps stretch most commonly used by athletes and nonathletes alike. Most people tend to have stronger but less flexible quadriceps muscles than hamstring muscles because of the tendency to stretch the hamstrings much more than the quadriceps. This creates an imbalance of strength and flexibility between the two muscle groups. To correct this imbalance, more emphasis needs to be placed on routinely stretching the quadriceps muscles.

When the right knee is extended behind the torso onto the floor, try to have a soft surface underneath the knee. This could be an exercise mat, grass, or even a pillow. This will minimize discomfort to the knee. When you move slowly into the stretched position, keep the left knee pointing forward. Do not let the left knee point to either side or let the right knee move along the surface of the floor. While the hips are placed in the forward direction, arching the back can increase the stretch on the muscles. This would stretch not only the quadriceps muscles but also the hip flexor muscles located in front of the pelvic area.

ADVANCED SUPPORTED STANDING KNEE EXTENSOR STRETCH

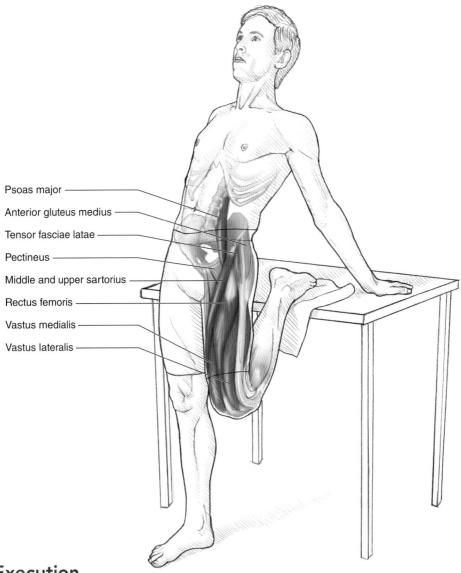

Psoas major

Anterior gluteus medius

Tensor fasciae latae

Pectineus

Middle and upper sartorius

Rectus femoris

Vastus medialis

Vastus lateralis

Execution

1. Stand with the back toward a padded table, bed, or soft platform that is below the height of the hips.
2. Balance your weight on the right leg, and bend the knee slightly.
3. Bend the left knee, and prop the left ankle on the rear support surface.
4. Place both hands on the rear support surface 6 to 12 inches (15 to 30 cm) behind the buttocks.
5. Move the torso back slowly so that the heel of the left foot touches the buttocks. Make sure the ankle and knee are comfortable.

6. Push the hips forward and simultaneously arch the back by bending the shoulders toward the buttocks.

7. Repeat this stretch for the opposite leg.

Muscles Stretched

Most-stretched muscles: Left vastus medialis, left vastus intermedius, left vastus lateralis, middle and upper right sartorius, left rectus femoris, left psoas major, left iliacus, left tensor fasciae latae

Less-stretched muscles: Left pectineus, anterior left gluteus medius

Stretch Notes

Knee stiffness can lead to injuries of the knee and of the quadriceps muscles and tendons. This is the most advanced stretch for the quadriceps muscles, and you must take extra care when attempting it. Because of the increased possibility of hyperflexing the knee, use this stretch only if you have very flexible muscles. By adhering to the following safety precautions, you can execute this stretch safely without injury.

While pulling the ankle slowly in a more backward than upward direction, concentrate on making sure your hips also move forward. This dual action stretches the hip flexor muscles located in front of the pelvic region as well as the quadriceps muscles. If you are experiencing soreness or tightness of either the lateral (outer) or medial (inner) side of the front thigh, consider placing most of the stretch emphasis on the medial muscles (vastus medialis and pectineus) by rotating the upper body away from the medial muscles (rotate the right side clockwise) when bending backward. To place most of the stretch emphasis on the lateral muscles (vastus lateralis and tensor fasciae latae), rotate the upper body away from the lateral muscles (rotate the right side counterclockwise) when bending backward.

For optimal results, it is important to brace both hands on the surface supporting the back. In addition, you should move your hips forward while carefully arching your back. This enables you to better control the amount of stretch being put on these muscles. Following these procedures maximizes stretch to the quadriceps muscles as well as to the hip flexor muscles located in front of the pelvic area. Yet another precaution for safety as well for comfort is having the ankle be up against the padded support behind you.

You might also consider moving the dorsal (top) part of the foot down to the padded support. This would bring additional benefits from the total stretch, because you also stretch the muscles in the anterior (front) part of the tibia bone in the lower leg. This is a powerful combination of multiple stretches.

In this stretch you are also able to change your trunk position, thus stretching the medial or lateral side of the thigh if you move your trunk in either a lateral (outer) or medial (inner) direction.

chapter 7

FEET AND CALVES

The skeletal structure of the lower leg and foot is made up of the long tibia and fibula bones found in the lower leg and the small foot bones called tarsals, metatarsals, and phalanges. These bones form numerous joints. The most important is the ankle joint, located between the tibia bone of the leg and the talus of the foot. This joint is a hinge joint, and it is involved with the major joint movements of plantar flexion (toes point down) and dorsiflexion (toes point up).

The other major joints found between each of the tarsal and metatarsal bones are gliding joints. They allow more limited movements of the foot. When several of these gliding joints are working together in the foot, a much broader range of movement is achieved compared with the movement produced by a single gliding joint working alone. Thus, multiple-joint movements allow for foot eversion (sole turned out) and inversion (sole turned in).

The joints that allow the most freedom of movement of the foot are the condyloid joints, located between the metatarsal bones and the phalanges. Condyloid joints allow the movements of flexion, extension, abduction, adduction, and circumduction of the toes. Finally, the joints that allow for flexion and extension of the toes are the hinge joints between the phalanges.

Without the ligaments and connective tissues found in the lower leg and foot, joint movement and muscle function would be greatly compromised. The joints in the foot are connected to each other by many ligaments. The largest ligament in this area is the deltoid ligament, or ankle medial collateral ligament. It is composed of four segments that connect the tibia to the talus, calcaneus, and navicular bones. Opposite the deltoid ligament is the ankle lateral collateral ligament, which is composed of three segments that connect the fibula to the talus and calcaneus bones. Since the deltoid ligament is much stronger than the ankle lateral collateral ligament, and the tibia is longer than the fibula, the ankle is predisposed toward inversion (turning in).

Retinacula are another type of connective tissue located in the lower leg that secure many of the muscle–tendon units. This support allows these muscles to work harder, stronger, and more efficiently. The superior and inferior retinacula in the dorsal (top) area of the foot hold down all the tendons of the extensor muscles. On the lower lateral side of the foot, the peroneal retinaculum holds down the tendons of the peroneus longus and peroneus brevis muscles. The flexor retinaculum on the medial side of the ankle holds down the tendons of the flexor digitorum longus, flexor hallucis longus, and posterior tibialis muscles.

The final noteworthy connective tissue is the plantar fascia. The plantar fascia is a broad, thick connective tissue that supports the arch on the bottom of the foot. It spans the area between the tuberosity of the calcaneus and the heads of the metatarsal bones.

The muscles that move the ankle and toes are located primarily in the lower leg (figure 7.1); these muscles have tendons that are as long as or longer than the muscles. The dominant tendon is the Achilles tendon, which is shared by the gastrocnemius, plantaris, and soleus. The gastrocnemius and soleus muscles are the prime plantar flexors and are assisted by the plantaris and tibialis posterior as well as two toe flexor muscles, flexor digitorum longus and flexor hallucis longus. Located on the outer (lateral) side of the calf is another group of three muscles—peroneus longus, peroneus brevis, and peroneus tertius—which are used in everting the foot. Additionally, the peroneus longus and peroneus brevis plantar flex the ankle.

Three anterior calf muscles (tibialis anterior, extensor hallucis longus, and extensor digitorum longus) dorsiflex the ankle as well as move the foot and toes. The extensor digitorum brevis, dorsal interosseous, and extensor hallucis brevis muscles are located on the dorsal (top) side of the foot and extend the toes. The muscles on the plantar (sole) side of the foot, the flexor digitorum brevis, quadratus plantae, flexor hallucis brevis, flexor digiti minimi, abductor hallucis, abductor digiti minimi, plantar interosseous, and lumbricales, are used to flex and spread the toes.

The movement range for the ankle and toes is limited by the strength of the agonist muscles, flexibility of the antagonist muscles, tightness of the ligaments, and bone contacts or impingements. One of the most notable limiters is the plantar fascia. A tight plantar fascia limits toe extension, and in cases where the fascia is inflamed, it will also limit plantar flexion. The range of motion for both plantar flexion and dorsiflexion can also be limited by the formation of bone spurs. Excessive stress can activate bone cells to form bone spurs on the anterior and posterior lips of the talus and the superior tibial dorsal neck. These bony outcroppings cause more rapid bone contacts, thus ending the movement. Interestingly, most of the range-of-motion limiters, except bone impingements, can be changed by doing stretching exercises.

On average, people are on their feet for a good part of the day. Therefore the muscles of the foot and lower leg are typically used more extensively during normal daily activities such as standing, walking, or running than any other muscles in the body. Although the musculature of the lower leg is substantially smaller than that of the upper leg, it supports the entire body and receives the heaviest load during these activities. Since the feet are also constantly exerting force against whatever surface they are in contact with, it is no wonder that many

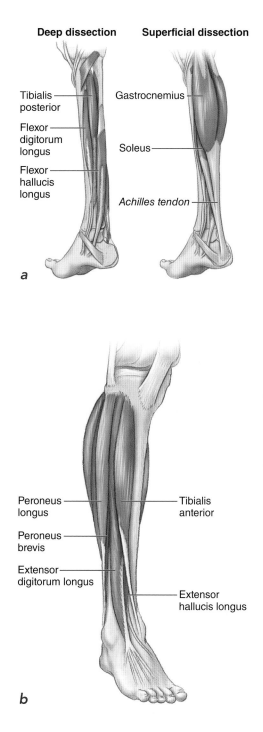

Figure 7.1 Calf and foot muscles: *(a)* posterior; *(b)* anterior.

people end up with minor aches, cramps, and weakness in the muscles of the lower legs and feet toward the end of the day. Stretching and strengthening these smaller muscle groups can alleviate much of the fatigue and pain caused by daily activities. In addition to helping reduce pain, stretching the muscles of the lower leg and foot can also improve overall flexibility, strength, strength endurance, balance, and stamina. Improving strength and flexibility in these muscle groups generally will enable a person to be more productive by increasing his ability to work longer and harder at work or during recreation activities.

Pain, cramping, restlessness, and weakness in the arch of the foot and calf muscles are common complaints. Problems such as these often result from the continual heavy loads put on the muscles. Chronic use of these muscles can also increase muscle tightness. Tightness may lead to conditions such as tendinitis and shin splints. Tendinitis of the Achilles tendon, associated with overuse and tightness of the gastrocnemius and soleus muscles, is quite common, in fact. Shin splints result from inflammation of the frontal compartment of the lower-leg muscles, the tibialis anterior and, in some cases, the soleus and flexor digitorum longus. Either of these conditions can become excruciating if not treated in the early stages. A variety of stretching and strengthening exercises for those muscle groups will, in most cases, improve these conditions and help prevent future episodes.

Another common condition is delayed-onset muscle soreness, or DOMS. This problem typically occurs after people participate in unusual or unfamiliar activities. The calf muscles tend to be affected by DOMS more often than any other muscle group in the body. Light stretching exercises are recommended to help improve this condition and relieve some of the pain associated with it.

As a rule, to effectively stretch specific muscles, the stretch must involve one or more movements in the opposite direction of the desired muscle's movements. For example, if you want to stretch the left flexor digitorum longus, perform a movement that involves dorsiflexion and eversion of the left ankle and toe extension of the left foot. When a muscle has a high level of stiffness, use fewer simultaneous opposite movements. For example, to stretch a very tight flexor digitorum longus, start by doing only toe extension. As the muscle becomes loose, incorporate more simultaneous opposite movements.

It is also recommended that the stretches in this chapter be tried from different angles of pull. Slightly altering the position of the body part changes the pull on the muscle. By altering positions, you can discover the location of any tight or sore spots within different muscle–tendon units. Also, if you change the position while stretching, you can add more versatility to your stretching program.

The leg and foot stretches in this chapter are grouped according to which muscle groups are being stretched. In addition, they are described in order from the easiest to the most difficult. Those who are new to a stretching program tend to be less flexible and should begin with the easiest level of stretches.

Progression to a more difficult stretch in this program should be made when the participant feels confident she is able to advance to the next level. For detailed instructions, refer to the stretching programs in chapter 9.

All the instructions and illustrations in this chapter are given for the right side of the body. Similar but opposite procedures are to be used for the left side. It should be noted that the stretches in this chapter are excellent stretches overall; however, not all of them may be completely suited to each person's specific needs.

BEGINNER SEATED TOE EXTENSOR STRETCH

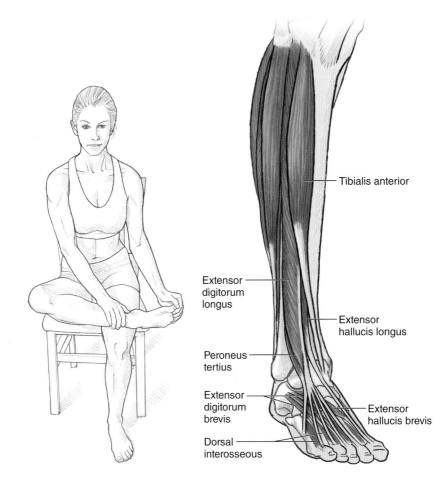

Tibialis anterior

Extensor digitorum longus

Extensor hallucis longus

Peroneus tertius

Extensor digitorum brevis

Extensor hallucis brevis

Dorsal interosseous

Execution

1. While sitting on a chair with the left foot on the floor, raise the right ankle and place it on top of the left knee.
2. While bracing the right ankle with the right hand, place the fingers of the left hand on the tops of the right toes.
3. Pull the tips of the toes toward the sole of the foot, away from the tibia bone.
4. Repeat this stretch for the opposite leg.

Muscles Stretched

Most-stretched muscles: Right extensor digitorum longus, right extensor digitorum brevis, right extensor hallucis longus, right extensor hallucis brevis, right tibialis anterior, right peroneus tertius

Less-stretched muscle: Right dorsal interosseous

Stretch Notes

This is a good stretch to alleviate minor aches and tightness in the toe extensor muscles located on the top of the foot. Generally speaking, these muscles are not as strong as the toe flexor muscles located on the bottom of the foot because they are not exerting force against the ground in daily activities such as running and walking. Rather, they are constantly being used as antagonist muscles in clearing the ground (toe extension and dorsiflexion) while walking or running. Consequently, they tend to become less sore or stiff than toe flexor muscles.

This stretch is one of the easiest to execute. It can be done at any time while you sit around, such as watching TV or doing any similar activity. When you are relaxing at the end of the day, regular stretching of these muscles will do wonders. A morning stretching routine is also a beneficial way to start the day. The series of stretching exercises can be done at any time during the day.

Hold the ankle firmly in order to keep it and the foot stable. You will feel the stretch on the top of the foot (dorsal side). If grasping and pulling on the tips of the toes causes too much pain, apply the pressure at the ball of the foot.

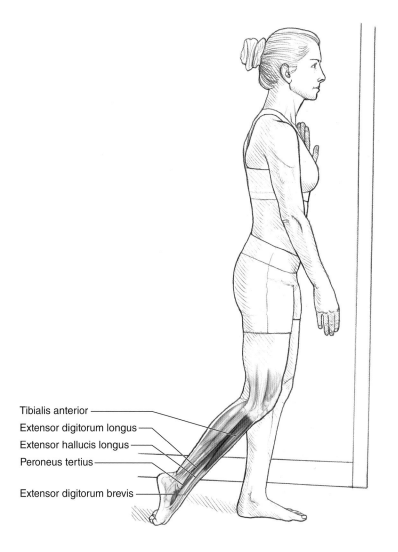

Tibialis anterior

Extensor digitorum longus

Extensor hallucis longus

Peroneus tertius

Extensor digitorum brevis

Execution

1. Stand upright and brace against a wall or an object for balance.

2. Point the right foot backward away from the body, with the dorsal (top) side of the toes down against the floor. Placing the top side of the foot on a pillow or towel makes this stretch more comfortable.

3. While keeping the dorsal side of the toes pressed against the floor, lean your weight onto the right leg, and press the bottom of the heel down toward the floor.

4. Repeat this stretch for the opposite leg.

Muscles Stretched

Most-stretched muscles: Right extensor digitorum longus, right extensor digitorum brevis, right extensor hallucis longus, right extensor hallucis brevis, right tibialis anterior, right peroneus tertius

Less-stretched muscle: Right dorsal interosseous

Stretch Notes

Many fitness exercisers experience shin splints in front of the tibia bone. This condition is very painful during exercise. This condition is associated with inflammation of the anterior tibialis muscle and the connective tissue around the anterior compartment of the tibia. It is often caused by overuse or tightness of the anterior tibialis muscle. It can also be associated with the type of shoes you wear and the surfaces where you exercise. People with shin splint problems definitely will benefit from this stretch. Shoes and running and walking surfaces also need to be evaluated.

It is more comfortable to perform this stretch on a carpet or other soft surface, or put a soft pillow or towel between the top part of the foot and the floor. Be sure not to drag the foot that is pressed to the floor. Moving the heel medially or laterally will place greater stretch on either the dorsal medial (inner) or dorsal lateral (outer) parts of the foot. It is also recommended that you explore each of these stretches from different angles of pull. This way you are able to locate the sore spots or tightness in these muscles. This stretch is more effective than the previous one. In this stretch your whole body weight puts more stress on these muscles as you stretch.

BEGINNER SEATED TOE FLEXOR STRETCH

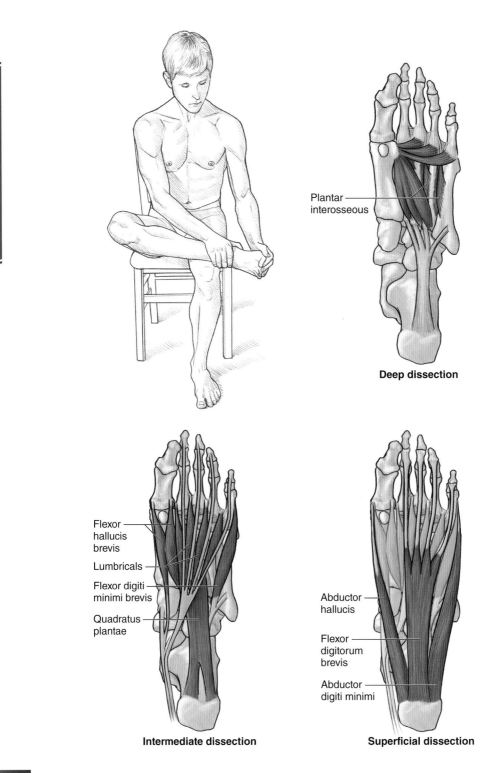

Plantar interosseous

Deep dissection

Flexor hallucis brevis

Lumbricals

Flexor digiti minimi brevis

Quadratus plantae

Intermediate dissection

Abductor hallucis

Flexor digitorum brevis

Abductor digiti minimi

Superficial dissection

Execution

1. While sitting on a chair with the left foot on the floor, raise the right ankle and place it on top of the left knee.
2. Brace the right ankle with the right hand, and place the fingers of the left hand along the bottoms of the toes of the right foot, with the fingers pointing in the same direction as the toes.
3. Use the fingers of the left hand to push the toes of the right foot toward the right knee.
4. Repeat this stretch for the opposite leg.

Muscles Stretched

Most-stretched muscles: Right flexor digitorum brevis, right quadratus plantae, right flexor digiti minimi brevis, right flexor hallucis brevis, right lumbricales, right plantar interosseous, right abductor hallucis, right abductor digiti minimi

Less-stretched muscles: Right flexor digitorum longus, right flexor hallucis longus, right tibialis posterior, right peroneus longus, right peroneus brevis, right plantaris, right soleus, right gastrocnemius

Stretch Notes

The foot muscles located in the arch of the foot receive constant stress during daily activities. The stress comes from supporting body weight during activities such as standing, walking, jumping, and running. Toe muscles also apply force against the ground whenever you are moving. Thus they are in constant use for most of the day, especially if you are an active person. After walking and standing long hours, foot muscles are often more tired, sore, and tight than any other muscle group in the body. You might even experience cramping of these muscles after a long day's work. Stretching these toe flexor muscles will help reduce the pain and soreness after a hard day's work and make you feel better. The muscles of the bottom of the foot are quite sensitive and respond to stretching exercises very well. Light massage along with light stretching exercises make you feel pleasantly relaxed after you have been on your feet most of the day.

Make sure to stabilize the foot and ankle with a firm hold. Pushing hard on the very ends of the toes with the left palm will provide a much greater stretch. You will feel the stretch on the sole (plantar side) of the foot.

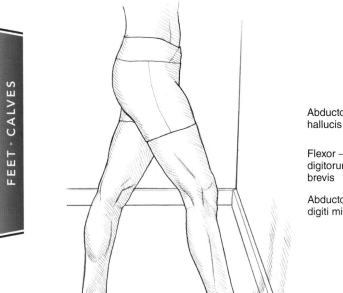

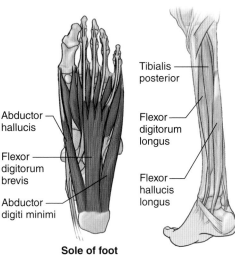

Tibialis posterior

Abductor hallucis

Flexor digitorum longus

Flexor digitorum brevis

Flexor hallucis longus

Abductor digiti minimi

Sole of foot

Execution

1. Stand upright while facing a wall, 1 to 2 feet (30 to 60 cm) away.

2. Keeping the heel of the right foot on the floor, press the bottoms of the toes of the right foot up against the wall. The ball of the foot should be more than half an inch (more than 2 cm) above the floor.

3. Lean forward and slide the ball of the right foot slowly down, keeping the toes pressed against the wall.

4. Repeat this stretch for the opposite leg.

Muscles Stretched

Most-stretched muscles: Right flexor digitorum brevis, right quadratus plantae, right flexor digiti minimi brevis, right flexor hallucis brevis, right lumbricales, right plantar interosseous, right abductor hallucis, right abductor digiti minimi

Less-stretched muscles: Right flexor digitorum longus, right flexor hallucis longus, right tibialis posterior

Stretch Notes

Have you driven a car for many hours without stopping? Have you ever felt that your foot is getting tired of moving the gas pedal up and down or being held in the same place for a long time? It happens to most of us. The muscles of the foot are not used to this. They simply get tired. This stretch or any of the previous stretches would be beneficial during a long drive.

Make sure the ball of the foot is parallel to the floor. This ensures that all the toes are stretched equally. Also, slide the ball of the foot down slowly. Otherwise, overstretching could happen. Bending the right knee slightly and moving the knee forward toward the wall will incorporate the calf muscles in the stretch.

BEGINNER PLANTAR FLEXOR STRETCH

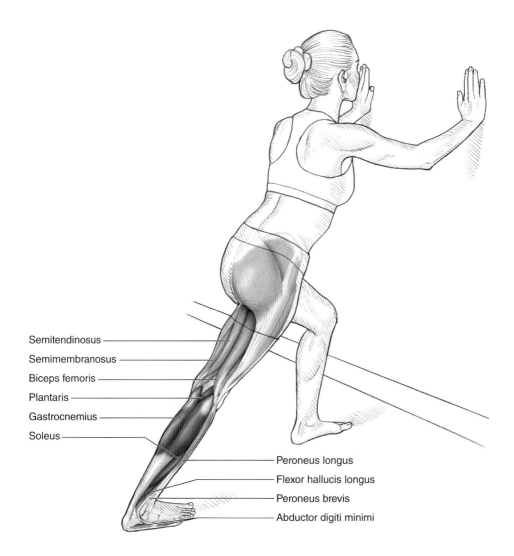

Semitendinosus
Semimembranosus
Biceps femoris
Plantaris
Gastrocnemius
Soleus

Peroneus longus
Flexor hallucis longus
Peroneus brevis
Abductor digiti minimi

Execution

1. Stand facing a wall, 2 feet (60 cm) away.

2. Brace your hands against the wall.

3. Keeping the left foot in place, place the right foot 1 to 2 feet (30 to 60 cm) behind the left foot. The left foot is 1 to 2 feet away from the wall, and the right foot is 2 to 4 feet (60 to 120 cm) away from the wall.

4. Keeping the right heel on the floor, lean your chest toward the wall. You can bend the left knee slightly to facilitate moving the chest up against the wall.

5. Repeat this stretch for the opposite leg.

Muscles Stretched

Most-stretched muscles: Right gastrocnemius, right soleus, right plantaris, right popliteus, right flexor digitorum longus, right flexor hallucis longus, right tibialis posterior

Less-stretched muscles: Right peroneus longus, right peroneus brevis, right flexor digitorum brevis, right quadratus plantae, right flexor digiti minimi brevis, right flexor hallucis brevis, right abductor digiti minimi, right abductor hallucis, right popliteus, right semitendinosus, right semimembranosus, right biceps femoris

Stretch Notes

Any time you start a new exercise program or participate in unusual or unfamiliar activities, you might experience muscular soreness during the days that follow. This is commonly known as delayed-onset muscle soreness, or DOMS. This painful feeling is felt most often 24 to 72 hours after the exercise. Walking or running uphill or downhill typically produces this painful effect. Calf muscles are usually affected more than any other muscle group in the body. Repeated stretching of these muscles for several days will help relieve the pain of DOMS.

As the chest gets closer to the wall, bending the knee slightly will realign the tibia and increase the distance between the muscle attachment points. This will increase the stretch on the tibialis posterior, flexor hallucis longus, and flexor digitorum longus muscles while at the same time reducing the stretch on the hamstring muscles.

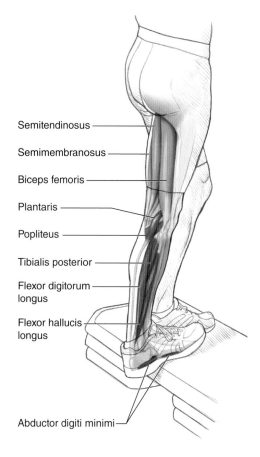

Semitendinosus

Semimembranosus

Biceps femoris

Plantaris

Popliteus

Tibialis posterior

Flexor digitorum longus

Flexor hallucis longus

Abductor digiti minimi

Execution

1. Stand upright on the edge of a stair or beam, with the midsection of the right foot on the edge. Hold onto a support with at least one hand.
2. Keep the right knee straight and the left knee slightly bent.
3. Lower the right heel as far as possible.
4. Repeat this stretch for the opposite leg.

Muscles Stretched

Most-stretched muscles: Right gastrocnemius, right soleus, right plantaris, right popliteus, right flexor digitorum longus, right flexor digitorum brevis, right flexor hallucis longus, right flexor hallucis brevis, right tibialis posterior, right quadratus plantae, right flexor digiti minimi brevis, right abductor digiti minimi, right abductor hallucis

Less-stretched muscles: Right semitendinosus, right semimembranosus, right biceps femoris

Stretch Notes

Many recreational and even competitive runners suffer from a condition called tendinitis, the chronic inflammation of a tendon. Tendinitis is caused by chronic overuse and tightness of the muscles associated with the affected tendons. The most vulnerable place for this condition in the lower leg is the Achilles tendon. The gastrocnemius and soleus muscles attach to this tendon. If not treated, tendinitis of the Achilles will become excruciatingly painful and will limit your participation in almost any sport activity. Research shows that most people simply don't spend enough time and effort stretching these muscles. Often it takes a long time (perhaps months) to get rid of this tendinitis. A good stretching program for these muscles should be part of your overall training program.

This stretch is the best for your calf muscles in general. It is more comfortable to do this stretch while wearing shoes. Always support the body—if the body is not supported, this could cause the muscles to contract and not stretch. After the heels reach their lowest point, apply more stretch by bending the knees slightly. This will stretch the tibialis posterior, flexor hallucis longus, and flexor digitorum longus muscles while reducing the stretch on the hamstring muscles. Placing the ball of the foot on the edge of the stairs or beam will increase the stretch on the origin (top part) of these muscle groups. Placing the midsection of the foot on the edge of the stairs or beam increases the stretch on the lower portion of these muscles. The sharper the edge of the stair, the better grip you are able to produce between the stair and the foot, and the more stretch you are able to produce on these muscles.

PLANTAR FLEXOR AND FOOT EVERTOR STRETCH

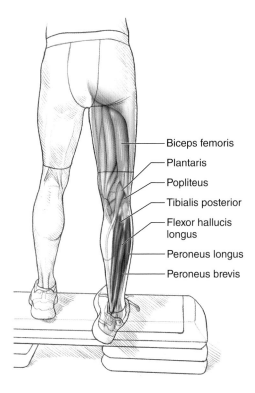

Biceps femoris
Plantaris
Popliteus
Tibialis posterior
Flexor hallucis longus
Peroneus longus
Peroneus brevis

Execution

1. Stand upright on the edge of a stair or beam, with the midsection of the right foot on the edge.
2. Place the foot in an inverted position by standing on the lateral (outer) side of the foot.
3. Keep the right knee straight and the left knee slightly bent.
4. Hold onto a support with at least one hand.
5. Keeping the foot inverted, lower the right heel as far as possible.
6. Repeat this stretch for the opposite leg.

Muscles Stretched

Most-stretched muscles: Right peroneus longus, right peroneus brevis, right peroneus tertius, right abductor digiti minimi, lateral side of right soleus, lateral side of right gastrocnemius, right flexor hallucis longus, right tibialis posterior

Less-stretched muscles: Right popliteus, right plantaris, medial head of right gastrocnemius, right biceps femoris, right flexor digitorum brevis, right quadratus plantae, right flexor digiti minimi brevis, right flexor hallucis brevis

Stretch Notes

Once in a while many of us experience soreness and tightness on the lateral (outer side) of the calf muscles. This could happen any time you walk or run on an uneven or unstable surface, such as grass or beach sand, or walk or run downhill. Often this soreness is felt on the days after the activity. This condition is called delayed-onset muscle soreness, or DOMS. When you encounter this problem, it is highly recommended to start a stretching program, especially of the muscles where this pain is felt. This particular stretch is helpful for the lateral (outer) side of the lower leg.

It is more comfortable to do this stretch while wearing shoes. This is an excellent stretch for the peroneus longus and peroneus brevis and the abductor digiti minimi muscles, which are located on the lateral (outer) side of the lower leg and the foot. Be extra careful when placing the foot in an inverted position, and be sure to progress slowly through this stretching exercise. After the right heel reaches the floor or the lowest point, increase the stretch by bending the right knee slightly. This removes any stretch on the hamstring muscles, but it further stretches the calf muscles.

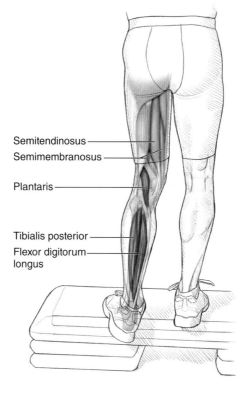

Semitendinosus

Semimembranosus

Plantaris

Tibialis posterior

Flexor digitorum longus

Execution

1. Stand upright on the edge of a stair or beam, with the midsection of the left foot on the edge.

2. Place the foot in an everted position by standing on the medial (inner) side of the foot.

3. Bend the left knee slightly toward the midsection of the body (inside direction), with the right knee slightly bent.

4. Hold on to a support with at least one hand.

5. While keeping the foot everted, lower the left heel as far as possible.

6. Repeat this stretch for the opposite leg.

Muscles Stretched

Most-stretched muscles: Left flexor digitorum longus, left abductor hallucis, medial side of left soleus, left tibialis posterior, left plantaris

Less-stretched muscles: Left flexor digitorum brevis, left quadratus plantae, left flexor hallucis brevis, left flexor digiti minimi brevis, left medial gastrocnemius, left semitendinosus, left semimembranosus

Stretch Notes

Shin splints are a nuisance for many endurance exercisers. This condition is often caused by overuse or tightness of the plantar flexor and invertor muscles. It is very hard to do any sport activities with the constant pain from shin splints. This stretch particularly stretches the flexor digitorum longus and the medial aspect of the soleus muscles. People with this problem will definitely benefit from this stretch. Also evaluate shoes as well as the running and walking surfaces. In addition, a thorough stretching program should be included in any rehabilitation program.

It is more comfortable to do this stretch while wearing shoes. This is an excellent stretch for the flexor digitorum longus, medial soleus, and abductor hallucis muscles, which are located on the medial side of the lower leg and foot. Take extra care when placing the foot in an everted position, and be sure to progress slowly through the stretch. After the left heel reaches the floor or the lowest point, bending the left knee slightly can increase the stretch. This reduces the stretch on the hamstring muscles, but it increases the stretch on the flexor digitorum longus, medial soleus, and abductor hallucis muscles.

chapter 8

DYNAMIC STRETCHES

Flexibility is an important component of physical fitness. Generally speaking, people with greater flexibility have better performances and reduced risk of injury. Consequently, many athletes include stretching exercises designed to enhance flexibility in both their training programs and preevent warm-up activities.

Since the late 1990s, however, several researchers have questioned the purported benefits of stretching. Numerous studies have established that preevent static stretching can inhibit almost all components of performance. For instance, preevent static stretching can reduce maximal strength, vertical jump performance, running speed, and muscular endurance. In addition, several recent research studies have failed to establish a link between preevent static stretching and injury prevention. In fact, a few studies have demonstrated that athletes with high levels of flexibility are more likely to suffer injuries than those with moderate flexibility. Some evidence shows that extremely tight people are less likely to experience muscle strains, but it is speculated that if preevent static stretching is reducing this type of injury, it is due to its ability to reduce the overall strength of the muscle. Strains, pulls, and tears happen when a muscle is forcefully contracted, so by reducing the force output you are less likely to cause injury. Finally, it is important to note that although many studies show the lack of benefits of preevent static stretching, there is still much evidence to support the benefits of static stretching after a workout.

Dynamic stretching is gaining popularity because of the complications that can arise from traditional preevent static stretching. As was discussed in the introduction, the muscle spindle proprioceptors have a fast dynamic component and a slow static component that provide information on not only the amount of length change but also the rate of length change. Fast length changes can trigger a stretch, or myotatic, reflex that attempts to resist the change in muscle length by causing the stretched muscle to contract. Slower stretches allow the muscle spindles to relax and adapt to new, longer lengths. Thus, dynamic activities that require quick, forceful movements, such as running, jumping, or kicking, utilize the dynamic receptor to limit flexibility. Consequently, researchers started to investigate whether a dynamic stretch that activates the dynamic receptor would be more beneficial when preparing to perform dynamic activities.

Dynamic stretching uses swinging, jumping, or exaggerated movements so that the momentum of the movement carries the limbs to or past the regular limits of the range of motion and activates a proprioceptive reflex response.

The proper activation of the proprioceptors can cause facilitation of the nerves that activated the muscle cells. This facilitation enables the nerves to fire more quickly, thus enabling the muscle to make fast and more powerful contractions. Hence, it can prepare the muscles and joints in a more specific manner since the body is going through motions it will likely repeat in the workout. It also helps the nervous system since dynamic motions do more to activate this aspect than static stretching does. Since dynamic stretching also includes constant motion throughout the warm-up, it maintains core body temperature, while static stretching can result in a drop in temperature of several degrees.

Research using dynamic stretching that controls movement through a joint's active range of motion has shown an increase in power performances such as sprinting and jumping. Moreover, there have been no reports of adverse effects on performance from either short or long sessions of dynamic stretching. For instance, one study showed that performance improved when dynamic stretches lasted more than 90 seconds, with little or no change for shorter stretch times. Additionally, a few research studies have shown that the negative impact of static stretching may be reduced or eliminated if dynamic stretches are done after static stretches. Thus, it is now highly recommended that a person perform dynamic stretches just before engaging in any activity.

As for any other activity, you must follow specific guidelines and principles when performing dynamic stretches:

- An effective warm-up that includes dynamic stretches should last 10 to 15 minutes or 10 to 20 repetitions.

- Observe your initial body position when you do a particular activity, and then make sure you start the dynamic stretch from the same initial position.

- Note the range of motion that each joint travels. The dynamic stretch should not greatly exceed the range of motion of the activity for which you are preparing. No bouncing.

- The dynamic stretch should closely replicate the movements used during the activity. Use good technique and be sure to use all of the muscles normally used during the activity. If the dynamic stretches mimic a specific sport skill such as a high knee lift, the stretches should utilize the specific factors of the skill. If you are careful to mimic the skill as closely as possible, you enhance the learning of the skill specificities and diminish the chances of introducing improper techniques.

- When doing dynamic stretches, you can either perform repetitions in the same place or travel for a set distance. Whether you remain in place or move, you should start each stretch slowly and progressively increase the range of motion and movement speed with each repetition. For instance, if you are moving over a distance, start with a walk, proceed to a skip, and finally end with a run.

- Dynamic stretches can be done singularly or in combination. Combining two or more stretches provides variety in your program and enables you to better mimic more complex skills.

In summary, each dynamic stretch should include 10 to 20 repetitions done either in place or over a given distance; you should progressively increase the range of motion and the speed of movement; the muscles should be contracted throughout the entire stretch; you should use good technique for each repetition just as you would normally perform the action; and you must ensure the movements are completely controlled by doing deliberate actions with no bouncing.

Those preparing for competitive or recreational activities can use the following dynamic stretches as a preexercise warm-up. In most cases, they are very helpful for almost any sport. These dynamic stretches concentrate on the major muscle groups in the body and are very easy to execute. You will find more enjoyment in your training or activity if you include these preexercise dynamic stretches in your program. In the next chapter, you will find more specific programs and recommendations for a variety of sporting events. You have multiple options from which to choose when deciding which stretching exercises best fit your purposes.

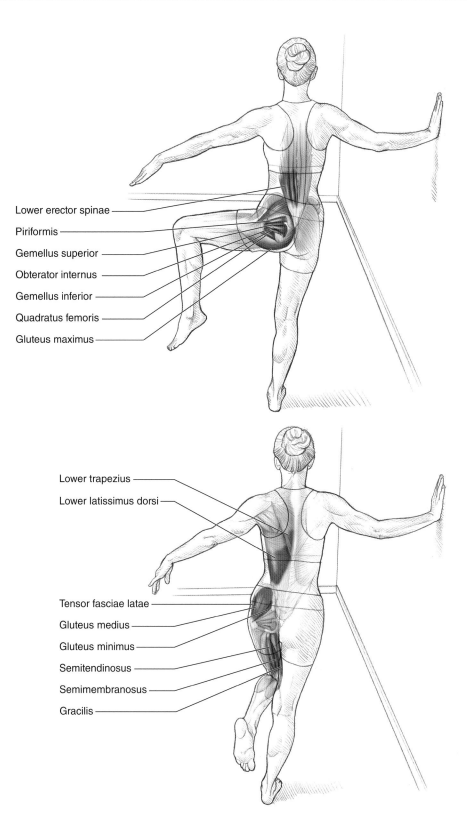

Lower erector spinae

Piriformis

Gemellus superior

Obterator internus

Gemellus inferior

Quadratus femoris

Gluteus maximus

Lower trapezius

Lower latissimus dorsi

Tensor fasciae latae

Gluteus medius

Gluteus minimus

Semitendinosus

Semimembranosus

Gracilis

Execution

1. Stand upright on the right leg, with the knee straight. Stand with the right side facing a supporting surface such as a wall, the edge of a corner, or a doorway. Hold on to the supporting object at shoulder height.
2. Bend the left knee and hip slightly, and let the left leg hang down in a relaxed manner as the starting point of this dynamic stretch.
3. Swing and rotate your left bent leg around your hip in a circular motion, inside and outside directions, in a dynamic manner.
4. Keep the trunk upright, and allow the circular motion to take place around the hip joint.
5. Repeat this stretch for the opposite leg.

Muscles Stretched

Most-stretched muscles in external rotation: Left gluteus maximus, left gluteus medius, left gluteus minimus, left piriformis, left gemellus superior, left gemellus inferior, left obturator externus, left obturator internus, left quadratus femoris, lower left erector spinae

Most-stretched muscles in internal rotation: Left gluteus medius, left gluteus minimus, left tensor fasciae latae, left semitendinosus, left semimembranosus, left gracilis, lower left latissimus dorsi, lower left trapezius

Stretch Notes

The hip external rotator muscles are located in the deep tissue of the hip just underneath the gluteus maximus muscle. These particular muscles can become sore or tight when unusual stress is placed on them or after engaging in activities that are not common in daily routines. Soreness or tightness is often due to extensive use of the hip external and internal rotator muscles in activities such as ice skating, in-line skating, or the skating style of cross-country skiing. Many other activities, such as an impromptu game of soccer requiring sprinting, jumping, and making sudden changes of direction, can easily result in uncomfortable or painful muscles later on.

On subsequent days, if soreness or tightness is still present in these particular muscles, before starting any activities that require hip external or internal rotational movements, use this dynamic stretch to warm up. This dynamic stretch increases the effectiveness of muscular movements and enhances total performance in many sporting activities.

DYNAMIC HIP ADDUCTOR AND ABDUCTOR STRETCH

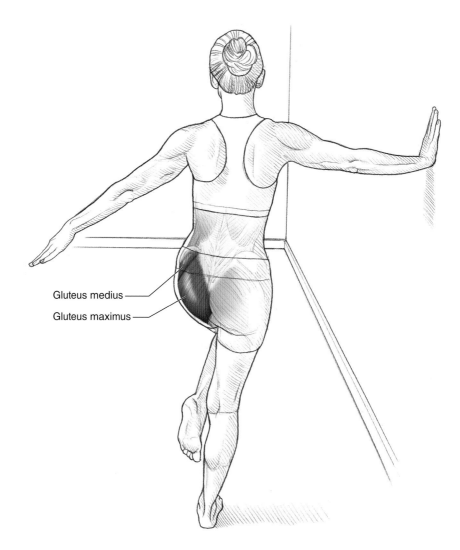

Gluteus medius
Gluteus maximus

Execution

1. Stand upright on the right leg, with the knee straight. Stand with the right side facing a supporting surface such as a wall, the edge of a corner, or a doorway. Hold on to the supporting object at shoulder height.

2. Bend the left knee and hip slightly, and let the left leg hang down in a relaxed manner as the starting point of this dynamic stretch.

3. Swing your left leg in front of you from side to side in a dynamic manner, with just enough clearance to avoid hitting the right leg. Make sure the knee of the swinging leg stays slightly bent.

4. Keep the trunk upright, and allow the movement to take place in the hip joint by using the adductor muscles located inside the thigh and hip and the abductor muscles located outside the thigh and hip.

5. Repeat this stretch for the opposite leg.

Muscles Stretched

Most-stretched muscles on inside of thigh: Left gracilis, left adductor magnus, left adductor longus, left adductor brevis, left pectineus, middle and lower left sartorius, left semitendinosus, left semimembranosus

Most-stretched muscles on outside of thigh: Left gluteus medius, left gluteus minimus, left gluteus maximus, left tensor fasciae latae, upper left sartorius

Stretch Notes

The muscles on the medial (inner) side and lateral (outer) side of the hip and thigh are fairly large. As a group, they are called the adductor and abductor muscles, respectively. These muscles are responsible for hip adduction (bringing the leg toward the midline of the body) and abduction (moving the leg away from the midline of the body). They also keep the legs centered under the body and are used as stabilizer muscles in the performance of daily activities. Certain unusual movements or activities, such as repeated stair climbing or hiking uphill or downhill, can cause the muscles in this region to feel sore or fatigued, a condition that could easily continue on subsequent days. Regular stretching most likely will alleviate some of the symptoms. It is strongly recommended that you stretch the adductor and abductor muscles both before and after participating in sports or other strenuous activities to help prevent injuries or symptoms.

This is a helpful and effective preexercise dynamic stretch for people who feel muscular pain or general stiffness in the inner or outer thigh. Pain in any region of the body is often a result of muscular soreness. When muscles are sore, they often feel stiff as well. A person with this condition has the tendency to limit the range of motion of the affected muscles in order to avoid pain. Therefore, normal daily activities can be significantly affected depending on the severity of the pain. Rather than avoiding movement, a person suffering from muscular soreness or tightness should specifically try to move and stretch the injured muscles in a dynamic manner before starting an exercise routine. Performing this dynamic stretch for the hip adductors and abductors will increase flexibility and warmth in these muscle groups just before the activity, which in turn will lessen the likelihood or severity of injury and also possibly increase exercise capacity.

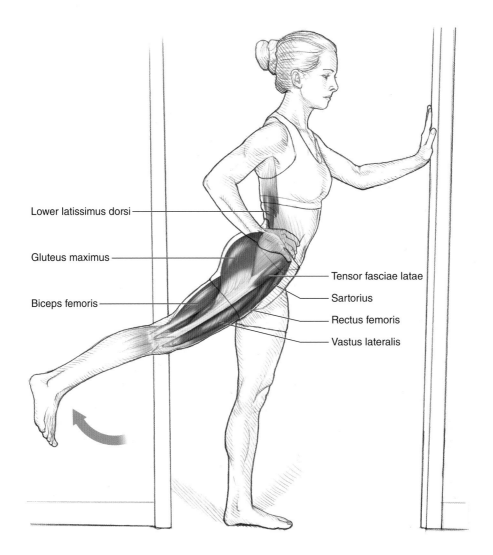

Lower latissimus dorsi

Gluteus maximus

Biceps femoris

Tensor fasciae latae

Sartorius

Rectus femoris

Vastus lateralis

Execution

1. Stand upright facing one side of a doorway. Stand on the left leg, with the knee straight. Hold on to the supporting object at the height of your shoulder.

2. Bend the right knee and hip slightly, and let the right leg hang down in a relaxed manner as the starting point of this dynamic stretch.

3. Keeping the leg slightly bent, swing your right leg straight forward and backward in a dynamic manner so that it swings parallel to the opening of the doorway.

4. Keep the trunk upright, and allow the movement to take place in front of and behind the hip joint, using the flexor and extensor muscles of the hip.

5. Repeat this stretch for the opposite leg.

Muscles Stretched

Most-stretched muscles on front of hip: Right rectus femoris, right vastus lateralis, right vastus intermedius, right vastus medialis, right tensor fasciae latae, right sartorius

Most-stretched muscles on back of hip: Right gluteus maximus, right semitendinosus, right semimembranosus, right biceps femoris, lower right erector spinae, lower right latissimus dorsi

Stretch Notes

The hip flexor and extensor muscles are used extensively in most sports. These muscles often fatigue first, and as a consequence performance decreases. Muscle soreness and tightness follow as the athlete continues to use these muscles. If they are not stretched properly, most likely the hamstrings and quadriceps will tighten up even more. Tight hamstrings and quadriceps are common among exercisers who significantly increase speed, the distance run, or the amount of uphill climbing during training. Tightness in the muscles can ease during exercise as the muscles get warmer, but when the athlete stops, the pain can return. Thus, it is especially important to stretch properly after exercise.

It is equally important to do some dynamic preexercise stretches before engaging in your regular exercise routine. This dynamic stretch for the hip flexors and extensors will alleviate some of the problems you might encounter as you exercise these muscles extensively. We recommend performing this stretch as a warm-up before doing any higher-intensity workouts.

DYNAMIC STRETCHES

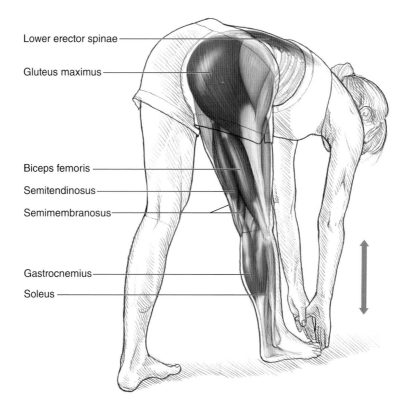

Lower erector spinae

Gluteus maximus

Biceps femoris

Semitendinosus

Semimembranosus

Gastrocnemius

Soleus

Execution

1. Stand upright with the right heel 1 to 2 feet (30 to 60 cm) ahead of the left toes.
2. Keeping the right knee straight and the left knee slightly bent, bend the trunk over toward the right knee.
3. Reach the hands toward the right foot.
4. Do the stretching in a dynamic manner by bobbing.
5. Repeat this stretch for the opposite leg.

Muscles Stretched

Most-stretched muscles: Right semitendinosus, right semimembranosus, right biceps femoris, right gluteus maximus, right gastrocnemius, lower right erector spinae

Less-stretched muscles: Right soleus, right plantaris, right popliteus, right flexor digitorum longus, right flexor hallucis longus, right tibialis posterior

Stretch Notes

When you start participating in a sport and do not stretch properly, you are more likely to have your hamstrings tighten up. Tight hamstrings are common among many athletes and those who participate in recreational activities. Tightness in these muscles can ease during exercise as the muscles get warmer, but when the athlete stops, the pain can return.

Tightness is often an indicator of minor or major muscle strains, a common occurrence mainly felt postexercise. In addition, muscle strength imbalances, in which the knee extensors are stronger or the gluteal muscles are weaker than the hamstrings, will also cause tightness. Thus, it is especially important to stretch properly after exercise because this is when the muscles are warm and more receptive to stretching.

This is the most common preexercise stretch for the hamstring and calf muscles. The hamstrings are used in most activities, and you might feel some discomfort in these muscles from your previous exercise session. In any type of fitness activity, minor aches and tightness in the hamstrings are possible. The optimal time to lightly stretch these muscles is just before you start another exercise session. In most cases, light dynamic stretches will relieve those uncomfortable symptoms, and you will feel so much better after engaging in these dynamic stretches.

For the best results, try to keep the right knee straight, and bend the torso directly from the hip. It is also important to keep the back as straight as possible. If you have tightness on the outer side of the hamstring muscles, turn the right foot slightly out and bend the head and trunk more toward the medial (inner) side of the right knee to increase the stretch of the biceps femoris. On the other hand, turning the right foot slightly in and bending the head and trunk more toward the lateral (outer) side of the knee will increase the stretch of the semitendinosus and semimembranosus muscles located on the inner side of the hamstring muscles.

DYNAMIC PLANTAR FLEXOR STRETCH

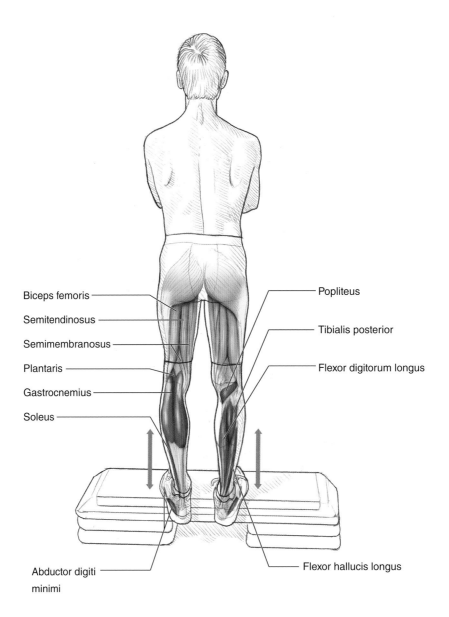

Biceps femoris

Semitendinosus

Semimembranosus

Plantaris

Gastrocnemius

Soleus

Popliteus

Tibialis posterior

Flexor digitorum longus

Flexor hallucis longus

Abductor digiti minimi

Execution

1. Stand upright on the edge of a stair or beam, with the midsection of both feet on the edge.
2. Hold on to a support with at least one hand, and keep the knees straight.
3. Lower the heels down as far as possible, and execute this stretch in a dynamic, bobbing manner.

Muscles Stretched

Most-stretched muscles: Gastrocnemius, soleus, plantaris, popliteus, flexor digitorum longus, flexor digitorum brevis, flexor hallucis longus, flexor hallucis brevis, tibialis posterior, quadratus plantae, flexor digiti minimi brevis, abductor digiti minimi, abductor hallucis

Less-stretched muscles: Semitendinosus, semimembranosus, biceps femoris

Stretch Notes

This stretch is often performed after exercise, but it is also highly recommended as a preexercise stretch. The calf muscles are in heavy use most of the day. They take most of the load during walking, running, and jumping activities. Naturally they become overworked, which sometimes leads to serious problems such as tendinitis or even muscle tears. As a preexercise stretch, this dynamic stretch for the plantar flexors will alleviate some of the problems you might encounter as you exercise these muscles extensively. We recommend performing this stretch as a warm-up before doing any higher-intensity workouts. You should also add a postexercise static plantar flexor stretch to your overall training program.

It is more comfortable to do this stretch while wearing shoes. Always support the body. If the body is not supported, this could cause the muscles to contract and not stretch. Do not overstretch these muscles when doing this exercise. Start easy, and slowly progress to a higher intensity level.

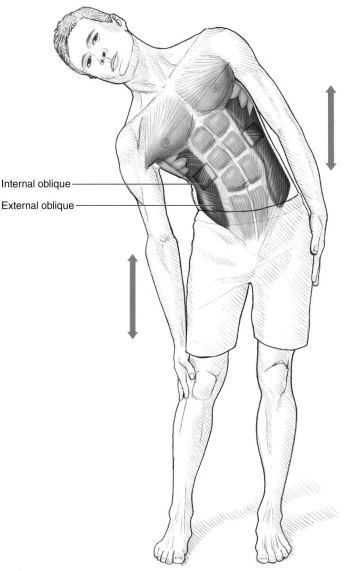

Internal oblique

External oblique

Execution

1. Stand upright with your feet shoulder-width apart.

2. Let the arms hang down by your sides.

3. With the help of your arms, bend your torso laterally back and forth in a dynamic manner. Move down and up your right side, with your right arm sliding down the right thigh toward your knee, followed by moving down and up on the left side, alternating between right and left.

4. Allow all dynamic movement to take place on the lateral side of the trunk.

Muscles Stretched

Most-stretched muscles: External oblique, internal oblique, intertransversarii, multifidus, quadratus lumborum, rotatores

Stretch Notes

Trunk lateral flexion stretching movements are often used in regular routines for nonspecific sports activity. You bend your trunk regularly in different directions many times a day. Most likely you might feel some unusual tightness or soreness in these muscles and simply want to get relief from these discomforts. Twisting the trunk goes along with lateral trunk flexion. These two muscle movements involve the muscles of the trunk extensors, flexors, and lateral flexors. Improved range of motion of all lower-trunk muscles can increase the range of motion in trunk lateral flexion and improve performance in activities that involve any nonspecific sports actions.

These core muscle groups are also often used as stabilizer muscles that allow other muscles to apply force. Thus it is important to keep these muscles in good shape. If these muscles are not working to their full capacity, it will affect the function of the other muscles, and your activity level and performance will naturally decrease.

It is important to warm up these muscles before performing any type of trunk flexion movement. Executing this stretch in a dynamic (ballistic) manner will definitely be helpful. This also decreases the possibility of injury or discomfort in these muscle groups during activity.

DYNAMIC TRUNK ROTATOR STRETCH

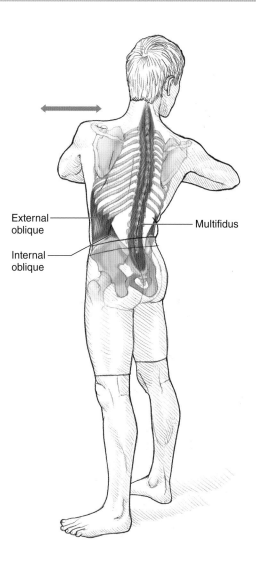

External oblique

Internal oblique

Multifidus

Execution

1. Stand upright with your feet shoulder-width apart; you can also do this stretch from a sitting position.

2. Bend your elbows and place your hands close to your chest. Keep your arms in this position during this stretch.

3. With the help of your arms, rotate your torso toward each side, back and forth in a dynamic manner.

4. Keep the trunk upright, and allow the dynamic movement to take place in the trunk.

Muscles Stretched

Most-stretched muscles: Multifidus, rotatores, external oblique, internal oblique

Stretch Notes

The trunk is considered a core area of the body. Trunk rotation is a very common movement in many sports as well as in common household activities. You bend your trunk regularly in daily activities, perhaps hundreds of times a day. No wonder you might encounter some muscular problems in this area. In addition, numerous sporting activities such as golf, tennis, and throwing sports require twisting of the trunk.

Twisting the trunk involves the trunk extensors, flexors, and lateral flexors. Improved range of motion of all lower-trunk muscles can increase the range of motion in trunk rotation and improve performance in activities that involve these actions. Warming up these muscles before any type of trunk rotation movement will definitely be helpful. Executing this stretch in a dynamic (ballistic) manner will also imitate the specific movement patterns experienced in these activities. This would decrease the possibility of injury or discomfort in these muscle groups during the activity.

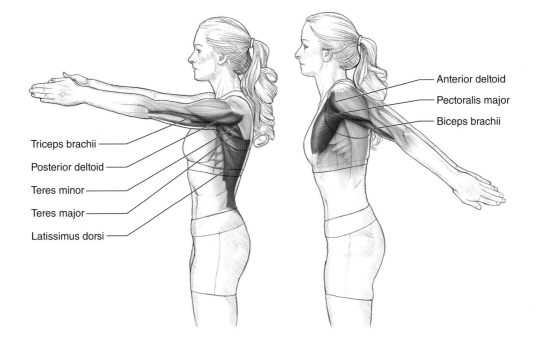

Anterior deltoid

Pectoralis major

Biceps brachii

Triceps brachii

Posterior deltoid

Teres minor

Teres major

Latissimus dorsi

Execution

1. Stand upright with your feet shoulder-width apart and your arms hanging down next to your hips.
2. Swing your arms forward and backward in a dynamic manner as far as you can through the full range of motion.
3. Keep the trunk upright, and allow the dynamic movement to take place in the shoulder joint.

Muscles Stretched

Most-stretched muscles in forward arm movement: Posterior deltoid, latissimus dorsi, teres major, teres minor, triceps brachii

Most-stretched muscles in backward arm movement: Biceps brachii, coracobrachialis, anterior deltoid, pectoralis major

DYNAMIC STRETCHES

Stretch Notes

You use these muscles extensively whenever you participate competitively or recreationally in any activities requiring under- or overhand throwing. Those who participate in these activities seasonally rather than all year long tend to encounter some tightness or soreness in the shoulder. This is a great preexercise stretch that should be performed whenever you have some tightness or aches in these muscles. This warm-up stretch is also a good way to loosen up your muscles to enhance the swinging patterns found in many sporting activities involving shoulder flexion and extension. This stretch imitates the dynamic movement patterns experienced during actual training sessions when throwing objects. Regularly stretch these muscles before and after these activities to prevent even further soreness and tightness. Continue to stretch these muscle groups as long as you participate in these activities. Warming up by using this dynamic stretching movement allows the body to get ready for your workout. This would decrease the possibility of injury or discomfort in these muscle groups.

DYNAMIC SHOULDER GIRDLE ABDUCTION AND ADDUCTION STRETCH

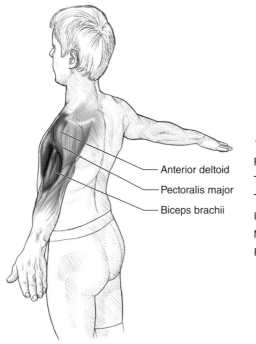

Anterior deltoid
Pectoralis major
Biceps brachii

Posterior deltoid
Teres minor
Teres major
Infraspinatus
Middle trapezius
Rhomboid

Execution

1. Stand upright with your feet shoulder-width apart.

2. Extend your arms out to the side, slightly lower than shoulder height.

3. Swing your arms laterally back and forth in front of the chest, bringing them in medially toward your body as far as you can so they cross over each other.

4. Keep the trunk upright, and allow the dynamic movement to take place in the shoulder joint.

Muscles Stretched

Most-stretched muscles during outward swing: Pectoralis major, pectoralis minor, anterior deltoid, coracobrachialis, biceps brachii

Most-stretched muscles during inward swing: Middle trapezius, rhomboids, posterior deltoid, teres major, teres minor, infraspinatus, supraspinatus

Stretch Notes

This stretch is a great preexercise movement for people who recreationally or competitively play any type of game that requires a racket, such as tennis, badminton, squash, and racquetball. This stretch relieves aches and tightness between the shoulder blades as well as in the chest. Also this is a good way to loosen up your swing patterns and bring smoothness to your performance. This stretch warms these muscles in order to get rid of any preexercise soreness or tightness, and it imitates the dynamic movement patterns experienced during training sessions. Warming up by using this dynamic stretching movement allows the body to get ready for your workout. It is always beneficial to perform a series of light stretches before starting any type of exercise, sport, or strenuous activity. These light stretches decrease the possibility of injury or discomfort in these muscle groups.

chapter 9

CUSTOMIZING YOUR STRETCHING PROGRAM

The programs in this chapter can be prescribed for anyone who is interested in improving flexibility, strength, and strength endurance. To make changes to any of these areas, you need to be involved in a regular stretching program, preferably as a daily routine or as close to that as possible. Changes will not come in a day or two but rather after a dedicated effort of several weeks. You can perform these programs with or without any other kind of exercise routine. According to the latest research, heavy stretching, even without any other exercise activity, can bring about changes in flexibility, strength, and muscular endurance.

As in any other exercise program, progression is an integral part of a successful stretching program. The stretching progression should be gradual, going from a lighter load with less time spent on each stretch to a heavier load with more time spent on each stretch. For these programs, begin with the initial program at the beginner's level and then progress through to the advanced level. You may customize this program according to your current level of experience and flexibility.

Generally, working through each level at the recommended speed will result in meaningful and consistent workouts. After such workouts, you will find improved flexibility in the muscles you worked as well as the satisfaction of having done something beneficial.

Intensity is always a critical factor when you want an exercise program to result in changes and improvements. In a stretching routine, intensity is controlled by the amount of pain associated with the stretch—in other words, how much effort you are putting into stretching the muscles. Using a pain scale from 0 to 10, initial pain is light (scale of 1 to 3) and usually dissipates as the stretching time is extended in each stretching routine. Light stretching occurs when you stretch a particular muscle group only to the point at which you feel the stretch, with an associated light pain. Moderate stretching (scale of 4 to 6) occurs when you start to feel increased, or medium, pain in the muscle you are stretching. In heavy stretching (scale of 7 to 10), you will initially experience a moderate to heavy pain at the start of the stretch, but the pain slowly dissipates as stretching continues.

Research studies have shown that heavier stretches rather than lighter stretches provide greater improvements in flexibility and strength. Thus, you are the key to your own success, and how well you are able to monitor stretch intensity and tolerate the pain level determines how quick and significant the improvements will be. Controlling the intensity is the key factor in any exercise program, and this also applies to your stretching program.

Because of the complexity of muscle attachments, many stretching exercises simultaneously affect a variety of muscle groups in the body and stretch the muscle groups around multiple joints. Thus, a small change in body position can change the nature of a stretch on any particular muscle. To get the maximal stretching benefit in any muscle, it is helpful to know the joint movements each muscle can do. Putting the joint through the full range of each motion allows for maximal stretching.

You can customize the stretches in this book, allowing for numerous stretch combinations. Also, this book illustrates only a portion of the available stretches. Experiment with these stretches by following the stretch notes. Information is also provided to enable you to explore a variety of positions in order to stretch the muscle by slightly altering the angles and directions of the various body positions. Thus, you can adapt the stretches to fit your individual needs and desires. For example, if you have soreness in only one of the muscles or just a part of the muscle, you can adapt each exercise to stretch that particular muscle. If the explained stretch or particular body position does not stretch a particular muscle as much as you want it to, then experiment by slightly altering the position. Keep making alterations in the position until you reach the desired level of stretch, based on the pain-scale rating.

In the programs that appear in this chapter, specific instructions are given relating to the time to hold the stretch and the time to rest between each stretch, as well as the number of repetitions you should do. Follow these instructions in order to get the benefits described. For example, if the instructions indicate you should hold a stretch position for 10 seconds, time or count out the stretch to ensure that you hold it for the recommended time. Also, you should incorporate only two to four heavier stretching days in each week and have a lighter stretching day in between each of the heavier stretching days.

It has been proven that stretching after weight training gives added benefits, especially when done either right after weight training or on nonlifting days. Stretching routines not only improve flexibility but also increase strength and strength endurance and improve balance. As we age, we all start losing some balance. Adding stretching exercises to your daily activities will bring additional improvements to your balance.

At the beginning of your stretching program, start each stretch with a light stretch and consider this your warm-up. After the initial warm-up stretch, move to your regular program. You need to slowly build up your tolerance to stretching and then move forward in the program as your flexibility improves. Tolerance is built by stretching on a regular basis, the same for any type of exercise program. Stretching is considered a workout just like any other exercise routine.

If your muscles seem tired, use only light stretches on those muscle groups. Do not overexert yourself. The body tells you if you need to back off. Keep in mind that the body needs to recover from any exercise routine, including any stretching routine. During the recovery period, the body builds up to a higher level. Chronic overuse of the muscles often leads to muscular fatigue, weakness, and even a partial failure of the muscle contraction.

Finally, for any stretch that requires you to sit or lie down, have a soft surface such as a carpet or athletic mat underneath you. Cushioning makes the exercises more comfortable and enjoyable. However, the cushioning should be firm. Too soft of a cushion will reduce the effectiveness of the stretch.

All the stretches in chapters 1 through 7 are best performed in a static manner by holding the stretch for a specified length of time. However, they also can be performed in a dynamic manner as a preexercise routine.

STATIC AND DYNAMIC STRETCHING PROGRAMS

The following programs are specific stretching recommendations based on initial flexibility. In addition to following the programs listed, you should follow these general recommendations:

1. Include all the major muscle groups of the body in your stretching program.
2. Do at least one stretch for each joint movement.
3. Before any physical activity, use only light stretches as part of the warm-up.
4. After an exercise routine, cool down with light- to medium-intensity stretches.
5. If muscles are sore after exercise, perform only light stretches two or three times, with a 5- to 10-second hold and 5 to 10 seconds of rest between each stretch.
6. If muscle soreness persists for several days, continue doing light stretches two or three times with a 5- to 10-second hold for each stretch.
7. The majority of the stretches should be static in nature.

Static Stretches

You get the most benefit when you do static stretches several times a week at the end of any other workout activity such as jogging or weightlifting. Depending on your initial level of flexibility, follow the guidelines and stretches detailed here and in tables 9.1 through 9.4.

Beginner Level

1. Hold the stretch position for 5 to 10 seconds.
2. Rest for 5 to 10 seconds between each stretch.
3. Repeat each stretch two or three times.
4. Use an intensity level on the scale from 1 to 3, with light pain.
5. Stretch for a total of 15 to 20 minutes each session.
6. Stretch two or three times per week.
7. Stay on this program at least four weeks before going to the next level.

Table 9.1   Beginner Static Stretch Routine

Area	Stretch	Page number
Neck	Neck extensor stretch	4
	Neck flexor stretch	8
Shoulders, back, and chest	Beginner shoulder flexor stretch	18
	Seated shoulder flexor, depressor, and retractor stretch	26
	Beginner shoulder extensor, adductor, and retractor stretch	28
Arms, wrists, and hands	Elbow flexor stretch	44
	Triceps brachii stretch	43
	Forearm pronator stretch with dumbbell	50
	Beginner wrist extensor stretch	54
	Beginner wrist flexor stretch	58
Lower trunk	Supine lower-trunk flexor stretch	76
	Seated lower-trunk extensor stretch	82
	Beginner lower-trunk lateral flexor stretch	86
Hips	Beginner seated hip external rotator stretch	96
	Hip and back extensor stretch	95
	Seated hip adductor and extensor stretch	110
Knees and thighs	Beginner seated knee flexor stretch	116
	Beginner seated knee extensor stretch	126
Feet and calves	Beginner seated toe extensor stretch	140
	Beginner seated toe flexor stretch	144
	Beginner plantar flexor stretch	148

Intermediate Level

1. Hold the stretching position for 15 to 20 seconds.
2. Rest for 15 to 20 seconds between each stretch.
3. Repeat each stretch three or four times.
4. Use an intensity level on the scale from 4 to 6, with moderate pain, two or three times per week.
5. Use an intensity level on the scale from 1 to 3 two or three times per week.
6. Stretch for a total of 30 to 40 minutes each session.
7. Stretch four or five times per week.
8. Stay on this program at least four weeks before going to the next level.

Table 9.2   Intermediate Static Stretch Routine

Area	Stretch	Page number
Neck	Neck extensor stretch	4
	Neck flexor stretch	8
Shoulders, back, and chest	Intermediate shoulder flexor stretch	20
	Seated shoulder flexor, depressor, and retractor stretch	26
	Intermediate shoulder extensor, adductor, and retractor stretch	30
	Shoulder adductor, protractor, and elevator stretch	32
Arms, wrists, and hands	Elbow flexor stretch	44
	Triceps brachii stretch	43
	Forearm pronator stretch with dumbbell	50
	Intermediate wrist extensor stretch	56
	Intermediate wrist flexor stretch	60
Lower trunk	Supine lower-trunk flexor stretch	76
	Intermediate lower-trunk lateral flexor stretch	88
Hips	Intermediate seated hip external rotator and extensor stretch	98
	Hip and back extensor stretch	95
	Seated hip adductor and extensor stretch	110
Knees and thighs	Intermediate standing knee flexor stretch	118
	Intermediate lying knee extensor stretch	128
Feet and calves	Beginner seated toe extensor stretch	140
	Beginner seated toe flexor stretch	144
	Beginner plantar flexor stretch	148

Advanced Level

1. Hold the stretch position for 25 to 30 seconds.
2. Rest for 25 to 30 seconds between each stretch.
3. Repeat each stretch five times.
4. Use an intensity level on the scale from 7 to 10, with heavy pain, two or three times per week.
5. Use an intensity level on the scale from 1 to 6 two or three times per week.
6. Stretch for a total of 50 to 60 minutes each session.
7. Stretch four or five times per week.
8. Stay on this level of the program for as long as you want.

Table 9.3   Advanced Static Stretch Routine

Area	Stretch	Page number
Neck	Neck extensor stretch	4
	Neck flexor stretch	8
Shoulders, back, and chest	Advanced shoulder flexor stretch	22
	Seated shoulder flexor, depressor, and retractor stretch	26
	Intermediate shoulder extensor, adductor, and retractor stretch	30
	Shoulder adductor, protractor, and elevator stretch	32
Arms, wrists, and hands	Elbow and wrist flexor stretch	46
	Triceps brachii stretch	43
	Intermediate wrist extensor stretch	56
Lower trunk	Prone lower-trunk flexor stretch	78
	Advanced standing lower-trunk lateral flexor stretch	90
Hips	Advanced standing hip external rotator stretch	100
	Hip and back extensor stretch	95
	Advanced seated hip adductor stretch	108
Knees and thighs	Advanced seated knee flexor stretch	120
	Advanced kneeling knee extensor stretch	130
Feet and calves	Advanced standing toe extensor stretch	142
	Advanced standing toe flexor stretch	146
	Advanced plantar flexor stretch	150
	Plantar flexor and foot evertor stretch	152

Expert Level

1. Hold the stretch position for 30 to 40 seconds.
2. Rest for 30 to 40 seconds between each stretch.
3. Repeat each stretch five times.
4. Use an intensity level on the scale from 7 to 10, with heavy pain, two or three times per week.
5. Stretch for a total of 50 to 60 minutes each session.
6. Stretch four or five times per week.
7. Stay on this level of the program for as long as you want.

Table 9.4   Expert Static Stretch Routine

Area	Stretch	Page number
Neck	Neck extensor stretch	4
	Neck flexor stretch	8
Shoulders, back, and chest	Assisted shoulder and elbow flexor stretch	24
	Assisted shoulder abductor stretch	36
Arms, wrists, and hands	Elbow and wrist flexor stretch	46
	Triceps brachii stretch	43
	Forearm pronator stretch with dumbbell	50
	Intermediate wrist extensor stretch	56
Lower trunk	Prone lower-trunk flexor stretch	78
	Advanced standing lower-trunk lateral flexor stretch	90
Hips	Advanced standing hip external rotator stretch	100
	Advanced seated hip adductor stretch	108
Knees and thighs	Expert raised-leg knee flexor stretch	122
	Advanced supported standing knee extensor stretch	132
Feet and calves	Advanced standing toe extensor stretch	142
	Advanced plantar flexor stretch	150
	Plantar flexor and foot evertor stretch	152

Dynamic Stretches (Preevent)

Ideally, dynamic stretches are done as part of a warm-up program just before engaging in an activity. Depending on your initial level of flexibility, it is recommended that you follow the suggested guidelines detailed here using all the stretches presented in chapter 8.

Beginner Level

1. Dynamically stretch using a bobbing motion for 5 to 10 seconds each stretch.
2. Rest for 5 to 10 seconds between each stretch.
3. Repeat each stretch two times.
4. Use an intensity level on the scale from 1 to 3, with light pain sensation.
5. Dynamically stretch for a total of 5 to 10 minutes each session.
6. Do these dynamic stretches as a warm-up procedure before engaging in an athletic event.
7. Stay on this program at least four weeks before going to the next level.

Intermediate Level

1. Dynamically stretch using a bobbing motion for 10 to 15 seconds each stretch.
2. Rest for 10 to 15 seconds between each stretch.
3. Repeat each stretch three times.
4. Use an intensity level on the scale from 1 to 3, with light pain sensation.
5. Dynamically stretch for a total of 10 to 15 minutes each session.
6. Do these dynamic stretches as a warm-up procedure before engaging in an athletic event.
7. Stay on this program at least four weeks before going to the next level.

Advanced Level

1. Dynamically stretch using a bobbing motion for 15 to 20 seconds each stretch.
2. Rest for 15 to 20 seconds between each stretch.
3. Repeat each stretch three times.
4. Use an intensity level on the scale from 4 to 7, with light pain sensation.
5. Dynamically stretch for a total of 15 to 20 minutes each session.
6. Do these dynamic stretches as a warm-up procedure before engaging in an athletic event.
7. Stay on this level of the program for as long as you want.

STRETCHING PROGRAM TO LOWER BLOOD GLUCOSE

In 2011, the *Journal of Physiotherapy* published a research study by Nelson, Kokkonen, and Arnall showing that a program of passive static stretches could lower blood glucose by an average of 18 percent after 20 minutes and 26 percent after 40 minutes. These researchers concluded that static stretching is an additional viable activity that can acutely help regulate blood glucose. Moreover, since stretching requires little effort, it appears to be an advantageous treatment for those with reduced physical capabilities. It can also be done without any additional equipment, facilities, or expenses and should easily fit into the repertoire of treatment modalities of any person with diabetes. In addition, since all the stretches in the study were done passively with the help of an assistant, if the person does the stretches actively by himself, the lowering effect on blood glucose should be greater.

The lowering of blood glucose by stretching relies on two major physiological principles. First, to enhance glucose transport out of the blood and into the muscles, the muscles should be held in the stretched position for at least 30 seconds. Second, holding a stretch for more than 30 seconds increases blood flow throughout the muscles, and improved blood flow is important for reducing blood glucose. Thus the stretching program detailed here and in table 9.5 is designed to first increase transport of glucose from the blood into the muscles and then to periodically enhance blood flow through each large muscle group.

Basic Guidelines

1. Hold the stretch position for 30 to 40 seconds each stretch.
2. Rest for 15 seconds between stretches.
3. Repeat each stretch four times.
4. Use an intensity level on the scale from 1 to 3, with light pain sensation.
5. Do all four stretches on one limb before doing the same stretches on the opposite limb.
6. Do the stretches in the order listed in table 9.5.

Table 9.5   Stretches That Lower Blood Glucose, In Order

Area	Stretch	Page number
Knees and thighs	Beginner seated knee flexor stretch	116
Hips	Seated hip adductor and extensor stretch	110
Shoulders, back, and chest	Advanced shoulder flexor stretch	22
Knees and thighs	Intermediate lying knee extensor stretch	128
Hips	Intermediate seated hip external rotator and extensor stretch	98
Shoulders, back, and chest	Intermediate shoulder extensor, adductor, and retractor stretch	30
Knees and thighs	Advanced seated knee flexor stretch	120
Feet and calves	Beginner plantar flexor stretch	148
Shoulders, back, and chest	Shoulder adductor and extensor stretch	34

SPORT-SPECIFIC STRETCHES

This section discusses recommended static stretches for those interested in developing or maintaining flexibility for 23 specific sports. An intermediate level of flexibility is the minimal requirement for doing these stretches. In addition to following the stretches listed, follow the general recommendations described in the section Static and Dynamic Stretching Programs as well as those listed for your specific level of flexibility.

Since dynamic stretches are better for preevent stretching, tables 9.6 through 9.28 include the dynamic stretches detailed in chapter 8 that will best benefit each sport.

Table 9.6 Stretches for Baseball, Position Player

	PREEVENT STRETCHES	
Area	**Stretch**	**Page number**
Shoulders, back, and chest	Dynamic shoulder flexion and extension stretch	174
	Dynamic shoulder girdle abduction and adduction stretch	176
Lower trunk	Dynamic trunk lateral flexion stretch	170
	Dynamic trunk rotator stretch	172
Hips	Dynamic hip external and internal rotator stretch	160
	Dynamic hip adductor and abductor stretch	162
	Dynamic hip flexor and extensor stretch	164
Knees and thighs	Dynamic standing knee flexor stretch	166
Feet and calves	Dynamic plantar flexor stretch	168
	TRAINING STRETCHES	
Area	**Stretch**	**Page number**
Shoulders, back, and chest	Intermediate shoulder flexor stretch	20
	Intermediate shoulder extensor, adductor, and retractor stretch	30
	Shoulder adductor, protractor, and elevator stretch	32
	Shoulder adductor and extensor stretch	34
Arms, wrists, and hands	Triceps brachii stretch	43
	Intermediate wrist extensor stretch	56
	Intermediate wrist flexor stretch	60
Lower trunk	Standing lower-trunk flexor stretch	80
	Intermediate lower-trunk lateral flexor stretch	88
Hips	Hip external rotator and back extensor stretch	104
	Hip and back extensor stretch	95
	Advanced seated hip adductor stretch	108
Knees and thighs	Advanced seated knee flexor stretch	120
	Advanced kneeling knee extensor stretch	130
Feet and calves	Plantar flexor and foot evertor stretch	152

Table 9.7 Stretches for Baseball, Pitcher

	PREEVENT STRETCHES	
Area	**Stretch**	**Page number**
Shoulders, back, and chest	Dynamic shoulder flexion and extension stretch	174
	Dynamic shoulder girdle abduction and adduction stretch	176
Lower trunk	Dynamic trunk lateral flexion stretch	170
	Dynamic trunk rotator stretch	172
Hips	Dynamic hip external and internal rotator stretch	160
	Dynamic hip adductor and abductor stretch	162
	Dynamic hip flexor and extensor stretch	164
Knees and thighs	Dynamic standing knee flexor stretch	166
Feet and calves	Dynamic plantar flexor stretch	168
	TRAINING STRETCHES	
Area	**Stretch**	**Page number**
Shoulders, back, and chest	Intermediate shoulder flexor stretch	20
	Shoulder adductor, protractor, and elevator stretch	32
	Shoulder adductor and extensor stretch	34
Arms, wrists, and hands	Elbow and wrist flexor stretch	46
	Triceps brachii stretch	43
	Intermediate wrist extensor stretch	56
	Intermediate wrist flexor stretch	60
Lower trunk	Standing lower-trunk flexor stretch	80
	Intermediate lower-trunk lateral flexor stretch	88
Hips	Hip external rotator and back extensor stretch	104
	Hip and back extensor stretch	95
	Advanced seated hip adductor stretch	108
Knees and thighs	Advanced seated knee flexor stretch	120
	Advanced kneeling knee extensor stretch	130
Feet and calves	Advanced plantar flexor stretch	150

Table 9.8 Stretches for Basketball

PREEVENT STRETCHES		
Area	**Stretch**	**Page number**
Shoulders, back, and chest	Dynamic shoulder flexion and extension stretch	174
	Dynamic shoulder girdle abduction and adduction stretch	176
Lower trunk	Dynamic trunk lateral flexion stretch	170
	Dynamic trunk rotator stretch	172
Hips	Dynamic hip external and internal rotator stretch	160
	Dynamic hip adductor and abductor stretch	162
	Dynamic hip flexor and extensor stretch	164
Knees and thighs	Dynamic standing knee flexor stretch	166
Feet and calves	Dynamic plantar flexor stretch	168
TRAINING STRETCHES		
Area	**Stretch**	**Page number**
Shoulders, back, and chest	Advanced shoulder flexor stretch	22
	Shoulder adductor, protractor, and elevator stretch	32
	Shoulder adductor and extensor stretch	34
Arms, wrists, and hands	Elbow and wrist flexor stretch	46
	Triceps brachii stretch	43
Lower trunk	Standing lower-trunk flexor stretch	80
	Seated lower-trunk extensor stretch	82
	Intermediate lower-trunk lateral flexor stretch	88
Hips	Hip external rotator and back extensor stretch	104
	Hip and back extensor stretch	95
	Advanced seated hip adductor stretch	108
Knees and thighs	Advanced seated knee flexor stretch	120
	Advanced kneeling knee extensor stretch	130
Feet and calves	Advanced standing toe flexor stretch	146
	Plantar flexor and foot evertor stretch	152

Table 9.9 Stretches for Bowling

	PREEVENT STRETCHES	
Area	**Stretch**	**Page number**
Shoulders, back, and chest	Dynamic shoulder flexion and extension stretch	174
	Dynamic shoulder girdle abduction and adduction stretch	176
Lower trunk	Dynamic trunk lateral flexion stretch	170
	Dynamic trunk rotator stretch	172
Hips	Dynamic hip external and internal rotator stretch	160
	Dynamic hip adductor and abductor stretch	162
	Dynamic hip flexor and extensor stretch	164
Knees and thighs	Dynamic standing knee flexor stretch	166
Feet and calves	Dynamic plantar flexor stretch	168
	TRAINING STRETCHES	
Area	**Stretch**	**Page number**
Shoulders, back, and chest	Advanced shoulder flexor stretch	22
	Intermediate shoulder extensor, adductor, and retractor stretch	30
	Shoulder adductor and extensor stretch	34
Arms, wrists, and hands	Intermediate wrist extensor stretch	56
	Intermediate wrist flexor stretch	60
	Wrist radial deviator stretch with dumbbell	62
	Wrist ulnar deviator stretch with dumbbell	64
Lower trunk	Standing lower-trunk flexor stretch	80
	Intermediate lower-trunk lateral flexor stretch	88
Hips	Advanced standing hip external rotator stretch	100
	Hip and back extensor stretch	95
	Advanced seated hip adductor stretch	108
Knees and thighs	Advanced seated knee flexor stretch	120
	Advanced kneeling knee extensor stretch	130
Feet and calves	Advanced plantar flexor stretch	150

Table 9.10 Stretches for Cycling

PREEVENT STRETCHES		
Area	**Stretch**	**Page number**
Shoulders, back, and chest	Dynamic shoulder flexion and extension stretch	174
	Dynamic shoulder girdle abduction and adduction stretch	176
Lower trunk	Dynamic trunk lateral flexion stretch	170
	Dynamic trunk rotator stretch	172
Hips	Dynamic hip external and internal rotator stretch	160
	Dynamic hip adductor and abductor stretch	162
	Dynamic hip flexor and extensor stretch	164
Knees and thighs	Dynamic standing knee flexor stretch	166
Feet and calves	Dynamic plantar flexor stretch	168
TRAINING STRETCHES		
Area	**Stretch**	**Page number**
Neck	Neck extensor stretch	4
	Neck flexor stretch	8
Shoulders, back, and chest	Intermediate shoulder extensor, adductor, and retractor stretch	30
	Shoulder adductor, protractor, and elevator stretch	32
Lower trunk	Standing lower-trunk flexor stretch	80
	Seated lower-trunk extensor stretch	82
	Intermediate lower-trunk lateral flexor stretch	88
Hips	Advanced standing hip external rotator stretch	100
	Hip external rotator and back extensor stretch	104
	Hip and back extensor stretch	95
	Advanced seated hip adductor stretch	108
Knees and thighs	Advanced seated knee flexor stretch	120
	Advanced kneeling knee extensor stretch	130
Feet and calves	Advanced standing toe extensor stretch	142
	Advanced plantar flexor stretch	150

Table 9.11 Stretches for Dance

PREEVENT STRETCHES		
Area	**Stretch**	**Page number**
Shoulders, back, and chest	Dynamic shoulder flexion and extension stretch	174
	Dynamic shoulder girdle abduction and adduction stretch	176
Lower trunk	Dynamic trunk lateral flexion stretch	170
	Dynamic trunk rotator stretch	172
Hips	Dynamic hip external and internal rotator stretch	160
	Dynamic hip adductor and abductor stretch	162
	Dynamic hip flexor and extensor stretch	164
Knees and thighs	Dynamic standing knee flexor stretch	166
Feet and calves	Dynamic plantar flexor stretch	168
TRAINING STRETCHES		
Area	**Stretch**	**Page number**
Neck	Neck extensor stretch	4
	Neck flexor stretch	8
Shoulders, back, and chest	Advanced shoulder flexor stretch	22
	Shoulder adductor and extensor stretch	34
Arms, wrists, and hands	Triceps brachii stretch	43
Lower trunk	Standing lower-trunk flexor stretch	80
	Intermediate lower-trunk lateral flexor stretch	88
Hips	Advanced standing hip external rotator stretch	100
	Hip and back extensor stretch	95
	Advanced seated hip adductor stretch	108
Knees and thighs	Advanced seated knee flexor stretch	120
	Advanced kneeling knee extensor stretch	130
Feet and calves	Advanced standing toe extensor stretch	142
	Advanced standing toe flexor stretch	146
	Advanced plantar flexor stretch	150

Table 9.12 Stretches for Diving

PREEVENT STRETCHES		
Area	**Stretch**	**Page number**
Shoulders, back, and chest	Dynamic shoulder flexion and extension stretch	174
	Dynamic shoulder girdle abduction and adduction stretch	176
Lower trunk	Dynamic trunk lateral flexion stretch	170
	Dynamic trunk rotator stretch	172
Hips	Dynamic hip external and internal rotator stretch	160
	Dynamic hip adductor and abductor stretch	162
	Dynamic hip flexor and extensor stretch	164
Knees and thighs	Dynamic standing knee flexor stretch	166
Feet and calves	Dynamic plantar flexor stretch	168
TRAINING STRETCHES		
Area	**Stretch**	**Page number**
Shoulders, back, and chest	Advanced shoulder flexor stretch	22
	Shoulder adductor and extensor stretch	34
	Assisted shoulder abductor stretch	36
Arms, wrists, and hands	Triceps brachii stretch	43
Lower trunk	Standing lower-trunk flexor stretch	80
	Seated lower-trunk extensor stretch	82
	Intermediate lower-trunk lateral flexor stretch	88
Hips	Advanced standing hip external rotator stretch	100
	Hip and back extensor stretch	95
	Advanced seated hip adductor stretch	108
Knees and thighs	Advanced seated knee flexor stretch	120
	Advanced kneeling knee extensor stretch	130
Feet and calves	Advanced standing toe extensor stretch	142
	Advanced standing toe flexor stretch	146
	Advanced plantar flexor stretch	150

Table 9.13 Stretches for American Football

Area	PREEVENT STRETCHES	
	Stretch	Page number
Shoulders, back, and chest	Dynamic shoulder flexion and extension stretch	174
	Dynamic shoulder girdle abduction and adduction stretch	176
Lower trunk	Dynamic trunk lateral flexion stretch	170
	Dynamic trunk rotator stretch	172
Hips	Dynamic hip external and internal rotator stretch	160
	Dynamic hip adductor and abductor stretch	162
	Dynamic hip flexor and extensor stretch	164
Knees and thighs	Dynamic standing knee flexor stretch	166
Feet and calves	Dynamic plantar flexor stretch	168
Area	TRAINING STRETCHES	
	Stretch	Page number
Neck	Neck extensor stretch	4
	Neck flexor stretch	8
Shoulders, back, and chest	Advanced shoulder flexor stretch	22
	Shoulder adductor and extensor stretch	34
Arms, wrists, and hands	Intermediate wrist extensor stretch	56
	Intermediate wrist flexor stretch	60
Lower trunk	Standing lower-trunk flexor stretch	80
	Intermediate lower-trunk lateral flexor stretch	88
Hips	Advanced standing hip external rotator stretch	100
	Hip and back extensor stretch	95
	Advanced seated hip adductor stretch	108
Knees and thighs	Advanced seated knee flexor stretch	120
	Advanced kneeling knee extensor stretch	130
Feet and calves	Advanced standing toe flexor stretch	146
	Plantar flexor and foot invertor stretch	154

Table 9.14 Stretches for Golf

	PREEVENT STRETCHES	
Area	**Stretch**	**Page number**
Shoulders, back, and chest	Dynamic shoulder flexion and extension stretch	174
	Dynamic shoulder girdle abduction and adduction stretch	176
Lower trunk	Dynamic trunk lateral flexion stretch	170
	Dynamic trunk rotator stretch	172
Hips	Dynamic hip external and internal rotator stretch	160
	Dynamic hip adductor and abductor stretch	162
	Dynamic hip flexor and extensor stretch	164
Knees and thighs	Dynamic standing knee flexor stretch	166
Feet and calves	Dynamic plantar flexor stretch	168
	TRAINING STRETCHES	
Area	**Stretch**	**Page number**
Shoulders, back, and chest	Advanced shoulder flexor stretch	22
	Intermediate shoulder extensor, adductor, and retractor stretch	30
	Shoulder adductor, protractor, and elevator stretch	32
	Shoulder adductor and extensor stretch	34
	Assisted shoulder abductor stretch	36
Arms, wrists, and hands	Intermediate wrist extensor stretch	56
	Intermediate wrist flexor stretch	60
Lower trunk	Standing lower-trunk flexor stretch	80
	Intermediate lower-trunk lateral flexor stretch	88
Hips	Hip external rotator and back extensor stretch	104
	Advanced seated hip adductor stretch	108
Knees and thighs	Advanced seated knee flexor stretch	120
	Advanced kneeling knee extensor stretch	130
Feet and calves	Advanced standing toe flexor stretch	146
	Advanced plantar flexor stretch	150

Table 9.15 Stretches for Gymnastics

PREEVENT STRETCHES		
Area	**Stretch**	**Page number**
Shoulders, back, and chest	Dynamic shoulder flexion and extension stretch	174
	Dynamic shoulder girdle abduction and adduction stretch	176
Lower trunk	Dynamic trunk lateral flexion stretch	170
	Dynamic trunk rotator stretch	172
Hips	Dynamic hip external and internal rotator stretch	160
	Dynamic hip adductor and abductor stretch	162
	Dynamic hip flexor and extensor stretch	164
Knees and thighs	Dynamic standing knee flexor stretch	166
Feet and calves	Dynamic plantar flexor stretch	168
TRAINING STRETCHES		
Area	**Stretch**	**Page number**
Neck	Neck extensor stretch	4
Shoulders, back, and chest	Advanced shoulder flexor stretch	22
	Intermediate shoulder extensor, adductor, and retractor stretch	30
	Shoulder adductor, protractor, and elevator stretch	32
	Shoulder adductor and extensor stretch	34
Arms, wrists, and hands	Elbow flexor stretch	44
	Triceps brachii stretch	43
Lower trunk	Standing lower-trunk flexor stretch	80
	Intermediate lower-trunk lateral flexor stretch	88
Hips	Hip and back extensor stretch	95
	Advanced seated hip adductor stretch	108
Knees and thighs	Advanced seated knee flexor stretch	120
	Advanced kneeling knee extensor stretch	130
Feet and calves	Advanced standing toe flexor stretch	146
	Advanced plantar flexor stretch	150

Table 9.16 Stretches for Handball and Racquetball

	PREEVENT STRETCHES	
Area	**Stretch**	**Page number**
Shoulders, back, and chest	Dynamic shoulder flexion and extension stretch	174
	Dynamic shoulder girdle abduction and adduction stretch	176
Lower trunk	Dynamic trunk lateral flexion stretch	170
	Dynamic trunk rotator stretch	172
Hips	Dynamic hip external and internal rotator stretch	160
	Dynamic hip adductor and abductor stretch	162
	Dynamic hip flexor and extensor stretch	164
Knees and thighs	Dynamic standing knee flexor stretch	166
Feet and calves	Dynamic plantar flexor stretch	168
	TRAINING STRETCHES	
Area	**Stretch**	**Page number**
Shoulders, back, and chest	Advanced shoulder flexor stretch	22
	Intermediate shoulder extensor, adductor, and retractor stretch	30
	Shoulder adductor, protractor, and elevator stretch	32
Arms, wrists, and hands	Elbow flexor stretch	44
	Triceps brachii stretch	43
Lower trunk	Standing lower-trunk flexor stretch	80
	Intermediate lower-trunk lateral flexor stretch	88
Hips	Hip external rotator and back extensor stretch	104
	Hip and back extensor stretch	95
	Advanced seated hip adductor stretch	108
Knees and thighs	Advanced seated knee flexor stretch	120
	Advanced kneeling knee extensor stretch	130
Feet and calves	Advanced standing toe extensor stretch	142
	Advanced standing toe flexor stretch	146
	Advanced plantar flexor stretch	150

Table 9.17 Stretches for Ice and Field Hockey

PREEVENT STRETCHES		
Area	**Stretch**	**Page number**
Shoulders, back, and chest	Dynamic shoulder flexion and extension stretch	174
	Dynamic shoulder girdle abduction and adduction stretch	176
Lower trunk	Dynamic trunk lateral flexion stretch	170
	Dynamic trunk rotator stretch	172
Hips	Dynamic hip external and internal rotator stretch	160
	Dynamic hip adductor and abductor stretch	162
	Dynamic hip flexor and extensor stretch	164
Knees and thighs	Dynamic standing knee flexor stretch	166
Feet and calves	Dynamic plantar flexor stretch	168
TRAINING STRETCHES		
Area	**Stretch**	**Page number**
Shoulders, back, and chest	Advanced shoulder flexor stretch	22
	Intermediate shoulder extensor, adductor, and retractor stretch	30
	Shoulder adductor, protractor, and elevator stretch	32
	Assisted shoulder abductor stretch	36
Arms, wrists, and hands	Elbow flexor stretch	44
	Triceps brachii stretch	43
Lower trunk	Standing lower-trunk flexor stretch	80
	Intermediate lower-trunk lateral flexor stretch	88
Hips	Hip external rotator and back extensor stretch	104
	Hip and back extensor stretch	95
	Advanced seated hip adductor stretch	108
Knees and thighs	Advanced seated knee flexor stretch	120
	Advanced kneeling knee extensor stretch	130
Feet and calves	Advanced standing toe extensor stretch	142
	Advanced plantar flexor stretch	150

Table 9.18 Stretches for Martial Arts

PREEVENT STRETCHES		
Area	**Stretch**	**Page number**
Shoulders, back, and chest	Dynamic shoulder flexion and extension stretch	174
	Dynamic shoulder girdle abduction and adduction stretch	176
Lower trunk	Dynamic trunk lateral flexion stretch	170
	Dynamic trunk rotator stretch	172
Hips	Dynamic hip external and internal rotator stretch	160
	Dynamic hip adductor and abductor stretch	162
	Dynamic hip flexor and extensor stretch	164
Knees and thighs	Dynamic standing knee flexor stretch	166
Feet and calves	Dynamic plantar flexor stretch	168
TRAINING STRETCHES		
Area	**Stretch**	**Page number**
Neck	Neck extensor stretch	4
Shoulders, back, and chest	Advanced shoulder flexor stretch	22
	Intermediate shoulder extensor, adductor, and retractor stretch	30
Arms, wrists, and hands	Intermediate wrist extensor stretch	56
	Intermediate wrist flexor stretch	60
Lower trunk	Standing lower-trunk flexor stretch	80
	Intermediate lower-trunk lateral flexor stretch	88
Hips	Advanced standing hip external rotator stretch	100
	Recumbent hip external rotator and extensor stretch	102
	Hip external rotator and back extensor stretch	104
	Advanced seated hip adductor stretch	108
	Seated hip adductor and extensor stretch	110
Knees and thighs	Advanced seated knee flexor stretch	120
	Advanced kneeling knee extensor stretch	130
Feet and calves	Advanced plantar flexor stretch	150

Table 9.19 Stretches for Running

	PREEVENT STRETCHES	
Area	**Stretch**	**Page number**
Shoulders, back, and chest	Dynamic shoulder flexion and extension stretch	174
	Dynamic shoulder girdle abduction and adduction stretch	176
Lower trunk	Dynamic trunk lateral flexion stretch	170
	Dynamic trunk rotator stretch	172
Hips	Dynamic hip external and internal rotator stretch	160
	Dynamic hip adductor and abductor stretch	162
	Dynamic hip flexor and extensor stretch	164
Knees and thighs	Dynamic standing knee flexor stretch	166
Feet and calves	Dynamic plantar flexor stretch	168
	TRAINING STRETCHES	
Area	**Stretch**	**Page number**
Shoulders, back, and chest	Advanced shoulder flexor stretch	22
	Intermediate shoulder extensor, adductor, and retractor stretch	30
Lower trunk	Standing lower-trunk flexor stretch	80
	Intermediate lower-trunk lateral flexor stretch	88
Hips	Advanced standing hip external rotator stretch	100
	Recumbent hip external rotator and extensor stretch	102
	Hip external rotator and back extensor stretch	104
	Hip and back extensor stretch	95
	Advanced seated hip adductor stretch	108
Knees and thighs	Advanced seated knee flexor stretch	120
	Advanced kneeling knee extensor stretch	130
Feet and calves	Advanced standing toe extensor stretch	142
	Advanced plantar flexor stretch	150
	Plantar flexor and foot evertor stretch	152
	Plantar flexor and foot invertor stretch	154

Table 9.20   Stretches for Snow Skiing

	PREEVENT STRETCHES	
Area	**Stretch**	**Page number**
Shoulders, back, and chest	Dynamic shoulder flexion and extension stretch	174
	Dynamic shoulder girdle abduction and adduction stretch	176
Lower trunk	Dynamic trunk lateral flexion stretch	170
	Dynamic trunk rotator stretch	172
Hips	Dynamic hip external and internal rotator stretch	160
	Dynamic hip adductor and abductor stretch	162
	Dynamic hip flexor and extensor stretch	164
Knees and thighs	Dynamic standing knee flexor stretch	166
Feet and calves	Dynamic plantar flexor stretch	168
	TRAINING STRETCHES	
Area	**Stretch**	**Page number**
Neck	Neck extensor stretch	4
Shoulders, back, and chest	Advanced shoulder flexor stretch	22
	Seated shoulder flexor, depressor, and retractor stretch	26
	Shoulder adductor and extensor stretch	34
Arms, wrists, and hands	Intermediate wrist extensor stretch	56
	Intermediate wrist flexor stretch	60
Lower trunk	Standing lower-trunk flexor stretch	80
	Intermediate lower-trunk lateral flexor stretch	88
Hips	Advanced standing hip external rotator stretch	100
	Hip external rotator and back extensor stretch	104
	Advanced seated hip adductor stretch	108
Knees and thighs	Advanced seated knee flexor stretch	120
	Advanced kneeling knee extensor stretch	130
Feet and calves	Advanced plantar flexor stretch	150
	Plantar flexor and foot invertor stretch	154

Table 9.21 Stretches for Soccer

PREEVENT STRETCHES		
Area	**Stretch**	**Page number**
Shoulders, back, and chest	Dynamic shoulder flexion and extension stretch	174
	Dynamic shoulder girdle abduction and adduction stretch	176
Lower trunk	Dynamic trunk lateral flexion stretch	170
	Dynamic trunk rotator stretch	172
Hips	Dynamic hip external and internal rotator stretch	160
	Dynamic hip adductor and abductor stretch	162
	Dynamic hip flexor and extensor stretch	164
Knees and thighs	Dynamic standing knee flexor stretch	166
Feet and calves	Dynamic plantar flexor stretch	168
TRAINING STRETCHES		
Area	**Stretch**	**Page number**
Shoulders, back, and chest	Advanced shoulder flexor stretch	22
	Seated shoulder flexor, depressor, and retractor stretch	26
	Shoulder adductor and extensor stretch	34
Lower trunk	Standing lower-trunk flexor stretch	80
	Intermediate lower-trunk lateral flexor stretch	88
Hips	Advanced standing hip external rotator stretch	100
	Hip external rotator and back extensor stretch	104
	Advanced seated hip adductor stretch	108
	Seated hip adductor and extensor stretch	110
Knees and thighs	Advanced seated knee flexor stretch	120
	Advanced kneeling knee extensor stretch	130
Feet and calves	Advanced standing toe extensor stretch	142
	Advanced standing toe flexor stretch	146
	Plantar flexor and foot evertor stretch	152
	Plantar flexor and foot invertor stretch	154

Table 9.22   Stretches for Swimming

PREEVENT STRETCHES		
Area	**Stretch**	**Page number**
Shoulders, back, and chest	Dynamic shoulder flexion and extension stretch	174
	Dynamic shoulder girdle abduction and adduction stretch	176
Lower trunk	Dynamic trunk lateral flexion stretch	170
	Dynamic trunk rotator stretch	172
Hips	Dynamic hip external and internal rotator stretch	160
	Dynamic hip adductor and abductor stretch	162
	Dynamic hip flexor and extensor stretch	164
Knees and thighs	Dynamic standing knee flexor stretch	166
Feet and calves	Dynamic plantar flexor stretch	168
TRAINING STRETCHES		
Area	**Stretch**	**Page number**
Shoulders, back, and chest	Advanced shoulder flexor stretch	22
	Assisted shoulder and elbow flexor stretch	24
	Seated shoulder flexor, depressor, and retractor stretch	26
	Intermediate shoulder extensor, adductor, and retractor stretch	30
	Shoulder adductor, protractor, and elevator stretch	32
	Shoulder adductor and extensor stretch	34
Arms, wrists, and hands	Triceps brachii stretch	43
Lower trunk	Standing lower-trunk flexor stretch	80
	Intermediate lower-trunk lateral flexor stretch	88
Hips	Hip external rotator and back extensor stretch	104
	Seated hip adductor and extensor stretch	110
Knees and thighs	Advanced seated knee flexor stretch	120
	Advanced kneeling knee extensor stretch	130
Feet and calves	Advanced standing toe extensor stretch	142
	Advanced plantar flexor stretch	150

Table 9.23　Stretches for Tennis

PREEVENT STRETCHES		
Area	**Stretch**	**Page number**
Shoulders, back, and chest	Dynamic shoulder flexion and extension stretch	174
	Dynamic shoulder girdle abduction and adduction stretch	176
Lower trunk	Dynamic trunk lateral flexion stretch	170
	Dynamic trunk rotator stretch	172
Hips	Dynamic hip external and internal rotator stretch	160
	Dynamic hip adductor and abductor stretch	162
	Dynamic hip flexor and extensor stretch	164
Knees and thighs	Dynamic standing knee flexor stretch	166
Feet and calves	Dynamic plantar flexor stretch	168
TRAINING STRETCHES		
Area	**Stretch**	**Page number**
Shoulders, back, and chest	Advanced shoulder flexor stretch	22
	Intermediate shoulder extensor, adductor, and retractor stretch	30
	Shoulder adductor, protractor, and elevator stretch	32
Arms, wrists, and hands	Elbow flexor stretch	44
	Triceps brachii stretch	43
	Intermediate wrist extensor stretch	56
Lower trunk	Intermediate lower-trunk lateral flexor stretch	88
Hips	Hip external rotator and back extensor stretch	104
	Hip and back extensor stretch	95
	Advanced seated hip adductor stretch	108
Knees and thighs	Advanced seated knee flexor stretch	120
	Advanced kneeling knee extensor stretch	130
Feet and calves	Advanced standing toe extensor stretch	142
	Advanced standing toe flexor stretch	146
	Advanced plantar flexor stretch	150

Table 9.24 Stretches for Track and Field, Sprints and Hurdles

	PREEVENT STRETCHES	
Area	**Stretch**	**Page number**
Lower trunk	Dynamic trunk lateral flexion stretch	170
	Dynamic trunk rotator stretch	172
Hips	Dynamic hip external and internal rotator stretch	160
	Dynamic hip adductor and abductor stretch	162
	Dynamic hip flexor and extensor stretch	164
Knees and thighs	Dynamic standing knee flexor stretch	166
Feet and calves	Dynamic plantar flexor stretch	168
	TRAINING STRETCHES	
Area	**Stretch**	**Page number**
Lower trunk	Standing lower-trunk flexor stretch	80
	Seated lower-trunk extensor stretch	82
	Intermediate lower-trunk lateral flexor stretch	88
Hips	Advanced standing hip external rotator stretch	100
	Hip external rotator and back extensor stretch	104
	Advanced seated hip adductor stretch	108
	Seated hip adductor and extensor stretch	110
Knees and thighs	Advanced seated knee flexor stretch	120
	Expert raised-leg knee flexor stretch	122
	Advanced kneeling knee extensor stretch	130
	Advanced supported standing knee extensor stretch	132
Feet and calves	Advanced standing toe extensor stretch	142
	Advanced standing toe flexor stretch	146
	Advanced plantar flexor stretch	150
	Plantar flexor and foot evertor stretch	152

Table 9.25 Stretches for Track and Field, Throwing Events

	PREEVENT STRETCHES	
Area	**Stretch**	**Page number**
Shoulders, back, and chest	Dynamic shoulder flexion and extension stretch	174
	Dynamic shoulder girdle abduction and adduction stretch	176
Lower trunk	Dynamic trunk lateral flexion stretch	170
	Dynamic trunk rotator stretch	172
Hips	Dynamic hip external and internal rotator stretch	160
	Dynamic hip adductor and abductor stretch	162
	Dynamic hip flexor and extensor stretch	164
Knees and thighs	Dynamic standing knee flexor stretch	166
Feet and calves	Dynamic plantar flexor stretch	168
	TRAINING STRETCHES	
Area	**Stretch**	**Page number**
Shoulders, back, and chest	Advanced shoulder flexor stretch	22
	Seated shoulder flexor, depressor, and retractor stretch	26
	Intermediate shoulder extensor, adductor, and retractor stretch	30
	Shoulder adductor and extensor stretch	34
Arms, wrists, and hands	Elbow and wrist flexor stretch	44
	Triceps brachii stretch	43
Lower trunk	Standing lower-trunk flexor stretch	80
	Intermediate lower-trunk lateral flexor stretch	88
Hips	Advanced standing hip external rotator stretch	100
	Hip external rotator and back extensor stretch	104
	Advanced seated hip adductor stretch	108
Knees and thighs	Advanced seated knee flexor stretch	120
	Advanced kneeling knee extensor stretch	130
Feet and calves	Advanced standing toe extensor stretch	142
	Advanced plantar flexor stretch	150

Table 9.26   Stretches for Volleyball

PREEVENT STRETCHES		
Area	**Stretch**	**Page number**
Shoulders, back, and chest	Dynamic shoulder flexion and extension stretch	174
	Dynamic shoulder girdle abduction and adduction stretch	176
Lower trunk	Dynamic trunk lateral flexion stretch	170
	Dynamic trunk rotator stretch	172
Hips	Dynamic hip external and internal rotator stretch	160
	Dynamic hip adductor and abductor stretch	162
	Dynamic hip flexor and extensor stretch	164
Knees and thighs	Dynamic standing knee flexor stretch	166
Feet and calves	Dynamic plantar flexor stretch	168
TRAINING STRETCHES		
Area	**Stretch**	**Page number**
Shoulders, back, and chest	Advanced shoulder flexor stretch	22
	Intermediate shoulder extensor, adductor, and retractor stretch	30
	Shoulder adductor and extensor stretch	34
Arms, wrists, and hands	Elbow and wrist flexor stretch	46
	Triceps brachii stretch	43
	Intermediate wrist flexor stretch	60
Lower trunk	Standing lower-trunk flexor stretch	80
	Intermediate lower-trunk lateral flexor stretch	88
Hips	Advanced standing hip external rotator stretch	100
	Hip external rotator and back extensor stretch	104
	Advanced seated hip adductor stretch	108
Knees and thighs	Advanced seated knee flexor stretch	120
	Advanced kneeling knee extensor stretch	130
Feet and calves	Advanced standing toe extensor stretch	142
	Advanced plantar flexor stretch	150

Table 9.27 Stretches for Weightlifting

PREEVENT STRETCHES		
Area	**Stretch**	**Page number**
Shoulders, back, and chest	Dynamic shoulder flexion and extension stretch	174
	Dynamic shoulder girdle abduction and adduction stretch	176
Lower trunk	Dynamic trunk lateral flexion stretch	170
	Dynamic trunk rotator stretch	172
Hips	Dynamic hip external and internal rotator stretch	160
	Dynamic hip adductor and abductor stretch	162
	Dynamic hip flexor and extensor stretch	164
Knees and thighs	Dynamic standing knee flexor stretch	166
Feet and calves	Dynamic plantar flexor stretch	168
TRAINING STRETCHES		
Area	**Stretch**	**Page number**
Neck	Neck extensor stretch	4
Shoulders, back, and chest	Advanced shoulder flexor stretch	22
	Intermediate shoulder extensor, adductor, and retractor stretch	30
	Shoulder adductor and extensor stretch	34
Arms, wrists, and hands	Elbow and wrist flexor stretch	46
	Triceps brachii stretch	43
	Intermediate wrist flexor stretch	60
Lower trunk	Standing lower-trunk flexor stretch	80
	Intermediate lower-trunk lateral flexor stretch	88
Hips	Advanced standing hip external rotator stretch	100
	Hip and back extensor stretch	95
	Advanced seated hip adductor stretch	108
Knees and thighs	Advanced seated knee flexor stretch	120
	Advanced kneeling knee extensor stretch	130
Feet and calves	Advanced standing toe flexor stretch	146

Table 9.28 Stretches for Wrestling

PREEVENT STRETCHES		
Area	**Stretch**	**Page number**
Shoulders, back, and chest	Dynamic shoulder flexion and extension stretch	174
	Dynamic shoulder girdle abduction and adduction stretch	176
Lower trunk	Dynamic trunk lateral flexion stretch	170
	Dynamic trunk rotator stretch	172
Hips	Dynamic hip external and internal rotator stretch	160
	Dynamic hip adductor and abductor stretch	162
	Dynamic hip flexor and extensor stretch	164
Knees and thighs	Dynamic standing knee flexor stretch	166
Feet and calves	Dynamic plantar flexor stretch	168
TRAINING STRETCHES		
Area	**Stretch**	**Page number**
Neck	Neck extensor stretch	4
	Neck flexor stretch	8
Shoulders, back, and chest	Advanced shoulder flexor stretch	22
	Shoulder adductor and extensor stretch	34
Arms, wrists, and hands	Elbow and wrist flexor stretch	46
	Triceps brachii stretch	43
Lower trunk	Standing lower-trunk flexor stretch	80
	Intermediate lower-trunk lateral flexor stretch	88
Hips	Advanced standing hip external rotator stretch	100
	Hip and back extensor stretch	95
	Advanced seated hip adductor stretch	108
Knees and thighs	Advanced seated knee flexor stretch	120
	Advanced kneeling knee extensor stretch	130
Feet and calves	Advanced standing toe extensor stretch	142
	Advanced standing toe flexor stretch	146

STRETCH FINDER

NECK

SHOULDERS, BACK, AND CHEST

ARMS, WRISTS, AND HANDS

FEET AND CALVES

DYNAMIC STRETCHES

ABOUT THE AUTHORS

Arnold G. Nelson, PhD, is a professor in the School of Kinesiology at Louisiana State University. A leading researcher on flexibility, he is considered one of the top authorities on the effect of stretching on muscle performance. Nelson is a fellow of the American College of Sports Medicine and earned his PhD in muscle physiology from the University of Texas at Austin. He resides in Baton Rouge, Louisiana.

Jouko Kokkonen, PhD, is a professor in exercise science at Brigham Young University in Hawaii. For more than two decades he has taught anatomy, kinesiology, exercise physiology, and athletic conditioning, and for more than three decades he has coached track and field. Kokkonen's research has focused on the acute and chronic effects of stretching. He earned his PhD in exercise physiology from Brigham Young University and now resides in Laie, Hawaii, with his wife, Ruthanne.

ANATOMY SERIES

Each book in the *Anatomy Series* provides detailed, full-color anatomical illustrations of the muscles in action and step-by-step instructions that detail perfect technique and form for each pose, exercise, movement, stretch, and stroke.

In the U.S. call **1-800-747-4457**
Australia 08 8372 0999 • Canada 1-800-465-7301
Europe +44 (0) 113 255 5665 • New Zealand 0800 222 062

HUMAN KINETICS
The Premier Publisher for Sports & Fitness
P.O. Box 5076, Champaign, IL 61825-5076 USA

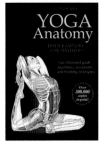